# How to Test Almost Everything Electronic

## 3rd Edition

*Delton T. Horn*

**TAB Books**
Division of McGraw-Hill

New York  San Francisco  Washington, D.C.  Auckland  Bogotá
Caracas  Lisbon  London  Madrid  Mexico City  Milan
Montreal  New Delhi  San Juan  Singapore
Sydney  Tokyo  Toronto

pbk    7   8   9   10   11   12   13   14   15   FGR/FGR   9   9   8   7   6
hc    1   2   3   4    5    6    7    8    9   10   FGR/FGR   9   9   8   7   6   5   4   3

**Library of Congress Cataloging-in-Publication Data**

Horn, Delton T.
     How to test almost everything electronic / Delton T. Horn. — 3rd
ed.
       p.     cm.
     Rev. ed. of: How to test almost everything electronic / Jack Darr
and Delton T. Horn.
     Includes index.
     ISBN 0-8306-4128-9 (h)     ISBN 0-8306-4127-0 (p)
     1. Electronic apparatus and appliances—Testing.   2. Electronic
measurements.   I. Darr, Jack.   How to test almost everything
electronic.   II. Title.
TK7878.H67   1993
621.381028'7—dc20                        93-12800
                                             CIP

Acquisitions editor: Roland S. Phelps
Editorial team: Suzanne Cheatle, Editor
                   Lori Flaherty, Managing Editor
                   Joann Woy, Indexer
Design team: Jaclyn J. Boone, Designer
                 Brian Allison, Associate Designer
Cover design: Sandra Blair, Harrisburg Pa.          EL2
Cover photograph: Brent Blair                  4227

# Contents

# Introduction

Electronics encompasses those things we can neither hear, see, touch, or taste. Electrical voltage, current, and resistance are invisible to the senses. Of course, you can directly sense an electrical shock, but to learn what's going on within an electronic circuit, you must use test equipment. *Test equipment* is special apparatus designed to give visible or audible indications of what is there and what it is doing.

Modern test instruments are amazingly versatile. They can do almost anything, if you know how to use them, how they work, and what their limitations are. If you don't know how to use test equipment properly, it is virtually useless. It is also just as important to know what a test instrument can't do. All test instruments can make the tests for which they are designed. Most can make many other tests as well, if you know how the instruments work and how their readings can indicate the presence or absence of other electrical parameters.

That's what this book is about—electronic tests and measurements, how to make them with all types of electronic test equipment, and how to interpret the results. Interpretation is the most important part of the whole process. It requires a full knowledge of both the test equipment and the circuits in which we're taking the readings.

Electronics is a rapidly growing field, with new developments appearing almost daily. Although the basic principles of electronics theory remain the same, often new technology calls for new test procedures. In this new third edition of *How to Test*

*Almost Everything Electronic*, every attempt has been made to provide as much up-to-date information as possible. Now-obsolete test procedures involving tube circuits were featured heavily in the original edition. This no-longer-relevant material has been eliminated and replaced with expanded information on testing transistor, and especially IC-based, circuitry.

The most important types of electronic test equipment are introduced in chapter 1. The following two chapters cover a wide variety of voltage and current tests, some of which are less than obvious.

The two most widely used test instruments are the multimeter (VOM, VTVM, or DMM) and the oscilloscope. Basic and advanced test procedures using these powerful test instruments are covered in chapters 4 and 5.

At some point in servicing an electronic circuit, you will need to determine whether or not a specific component has gone bad. Chapter 6 offers many tips on testing various specific types of electronic components.

A number of test procedures for television circuits are given in chapter 7. Chapter 8 features a number of more or less unclassifiable special test procedures, including tests for various types of semiconductors. Chapter 9 covers signal tracing and alignment tests.

Chapter 10, on digital circuits, has been significantly expanded in this edition in response to the growing emphasis on digital circuitry in modern commercial equipment of all types.

Finally, chapter 11 covers the important principles of flowcharting and troubleshooting complex systems comprised of multiple circuits. This information will help you pinpoint the problematic stage, saving a lot of time that could be wasted in making unnecessary tests. Devoting a little time to thinking about the circuit before you turn on your test equipment will never be wasted time. In fact, in the long run, most servicing jobs will go much faster.

As you will discover as you go through this book, there are many shortcuts—combinations of instruments and so forth—that you can use to test almost any electrical quantity, even those that might not appear to be within the range of your test equipment. A surprising number of electronic tests can be made with little more than a pilot lamp, a neon lamp, or a dc voltmeter. However, other tests require more sophisticated, and expensive, equipment.

Although it is obviously impossible for any single volume to cover every possible type of electronic test procedure, our goal has been to offer as wide a range of generally applicable procedures as possible. In updating the material for this book, we have tried to live up to the title as much as possible and help the reader learn "how to test almost everything electronic."

# 1
# Test equipment

IN SOME CASES, YOU CAN DIAGNOSE A PROBLEM SIMPLY BY EXAMINING the symptoms logically. More often, however, any given symptom or set of symptoms indicates several possible faults. The service technician depends on his test equipment. Although certainly some technicians are better than others, to an extent no technician is better than his or her test equipment. The best technician in the world would be severely limited with inadequate test equipment.

Various common types of test equipment are mentioned throughout this book. In this section, we specifically focus on several different types of test equipment, emphasizing fairly recent devices as much as possible.

The first requirement for a well-equipped electronics workbench is a large work surface. You might find yourself with several pieces of equipment opened up and spread out at the same time, plus you need room for any test equipment. Therefore, your work surface should be as large as possible.

Adequate lighting is an absolute must. A dimly lit work area will slow you down and increase the chance of errors. Use multiple light sources so you can't block off the light with your own body or a piece of equipment.

A high-intensity lamp on a flexible goose-neck, as illustrated in Fig. 1-1, can be an extremely handy item to have—you can focus the light wherever you need it. A lot of equipment has dark nooks and crannies you need to see into.

Your work area should have several electrical outlets. You can never have too many. There always seems to be one more

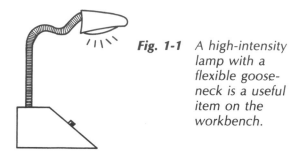

**Fig. 1-1** *A high-intensity lamp with a flexible goose-neck is a useful item on the workbench.*

thing you need to plug in. Avoid using cube taps and extension cords as much as possible because they can be fire hazards if overloaded. Many distributors sell power strips, which are convenient and safe (Fig. 1-2). Many even have built-in surge protectors. Most have a handy master power switch that is useful in some circumstances. It is also a good way to ensure everything is off when you close up for the night.

What test equipment will you need? It depends on just what type of work you do. A multimeter (VOM or VTVM) is absolutely essential for electronics workbenches. If possible, you should have at least one VOM and one VTVM. The multimeter is the technician's right arm.

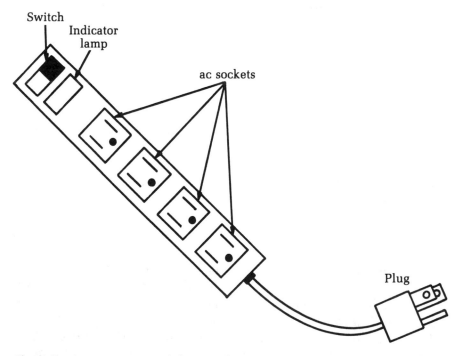

**Fig. 1-2** *A power strip safely provides extra ac outlets.*

The technician's left arm is the oscilloscope, another essential device for virtually all electronic workbenches. Anytime you are dealing with an ac signal, a scope is useful because it permits you to actually see the waveform. You also can measure the voltage at any point in the cycle, check for distortion, measure the cycle period, and make many other tests.

The multimeter and the oscilloscope are standard items used in all types of electronics work. Other generally useful devices include variable power supplies, Variacs, and signal generators. There are many different types of signal generators for various applications; several common types are discussed in this chapter. Other devices include capacitance meters, frequency counters, specialized signal generators, and signal tracers.

Don't scrimp on your test equipment. Your accuracy will be no better than the capabilities of your equipment. Don't overspend on features you don't need or on equipment you will rarely, if ever, use. Make sure, however, that you do get what you need. Nothing is worse than being stumped on a servicing job because you don't have the right equipment. Occasionally, this situation is inevitable because you will undoubtedly run into something outside your usual area. If insufficient equipment problems plague you frequently, however, you need to be better prepared.

## Multimeters

A *multimeter* measures various electrical parameters — usually voltage, current, and resistance. Multimeters are used in many of the test procedures in this book, especially in chapters 2 and 4.

A VOM (voltohm-milliammeter) is a passive device. It does not include an active amplifier or buffer stage, so the input impedance (resistance) is fairly low. Some cheap VOMs have input impedances as low as 1,000 ohms ($\Omega$) per volt. Such units are virtually useless for practical electronics work because they can excessively load down the circuit being tested, throwing off the measurement.

The unofficial standard for professional VOMs has long been considered 20,000 $\Omega$ per volt. Modern, high-quality VOMs usually have input impedances of 50,000 or 100,000 $\Omega$ per volt.

VOMs are usually lightweight and portable. Most VOMs have a small internal battery for the ohmmeter section, which means you don't have to be tied to a power source.

A much higher meter impedance is necessary for some precision measurements. In this case, use a multimeter with an active amplifier/buffer stage. In the past, the amplifier was a tube circuit, so this type of device traditionally has been known as a vacuum tube voltmeter (VTVM). Today, of course, tubes are increasingly scarce. Even when such a device uses only semiconductor components, it is still often called a VTVM. A better name might be electronic multimeter (EMM).

VTVMs, or EMMs, have high input impedances, typically 1 MΩ per volt. They are very precise, but tend to be more expensive and bulkier than VOMs. Also, a VTVM or EMM, being an active device, requires some sort of power supply.

Today, the trend is toward digital. Digital multimeters, or DMMs, are extremely popular. Many are as small and portable as a VOM. Often, they are powered from a small internal battery (9-volt transistor radio batteries are often used). A DMM usually has an input impedance comparable to a VTVM, but it also offers the convenience of a lightweight VOM.

Digital multimeters offer a number of significant advantages over their analog counterparts. They are easier to read without ambiguity. A display reading of 7.83 volts (V) is quite unmistakable and precise. On an analog VOM, you must approximate the value from the position of the meter's pointer between standardized mark-points. It also can be difficult to determine exactly where the pointer actually is in relation to the meter's scale face, especially if viewed from any angle other than straight head-on. Parallex errors are practically inevitable when you are using an analog meter. Parallex errors are not a problem with a DMM, regardless of the viewing angle. Either you can read the displayed numbers or you can't. If you look at a DMM from a bad angle, you are not going to mistake a reading of 7.83 for 7.69 or 8.12.

Another advantage of DMMs is their high input impedances, typically at least 1 MΩ (1,000,000 Ω) per volt. A high input impedance translates to high accuracy and sensitivity. In an analog VOM, the input impedance is usually much lower. For a long time, 20 k (20,000 Ω) was considered standard, although many 50 k (50,000 Ω) and 100 k (100,000 Ω) VOMs are available. A VOM with a FET input stage, or a VTVM, generally has a higher input impedance than a standard VOM. Some inexpensive VOMs have input impedances as low as 2 k (2,000 Ω) per volt. Such units are not suitable for serious electronics work.

The importance of a high input impedance for a voltmeter is illustrated in Fig. 1-3. The voltmeter measures the voltage drop across some resistive element in the circuit, but the internal resistance (impedance) of the voltmeter itself is in parallel with the

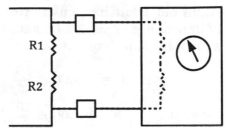

**Fig. 1-3** *The input impedance of a voltmeter acts like a parallel resistance across the circuit under test, affecting the reading.*

intended circuit resistance. This situation can affect the resulting reading, often to a surprising degree.

Let's assume we are feeding 9 V through two resistors, as shown in the diagram. Resistor R1 had a value of 25 k (25,000 $\Omega$) and resistor R2's value is 50 k (50,000 $\Omega$). The total series resistance is simply the sum of the two component resistances.

$$R_t = R_1 + R_2$$
$$= 25,000 + 50,000$$
$$= 75,000 \ \Omega$$
$$= 75 \ k$$

The current flow through the resistors can be found with Ohm's law:

$$I = \frac{E}{R}$$

$$= \frac{9}{75,000}$$
$$= 0.00012 \ A$$
$$= 0.12 \ mA$$

The same amount of current flows through each resistor in series, so the voltage drop across an individual resistor ($R^2$, in our example) can be found be rearranging the Ohm's law equation.

$$E = IR$$
$$= 0.00012 \times 50,000$$
$$= 6 \ V$$

This is all quite simple and straightforward. When you try to actually measure this voltage drop, however, the input impedance of the multimeter is added in parallel with resistor R2. The effective total resistance in the circuit becomes equal to:

$$R_t = R_1 + \frac{R_2 \times R_z}{R_2 + R_m}$$

Consider what happens if you use a cheap VOM with an input impedance of just 2 k (2,000 Ω). The circuit's total effective resistance becomes equal to:

$$R_t = 25,000 + \frac{50,000 \times 2,000}{50,000 + 2,000}$$
$$= 25,000 + \frac{100,000,000}{52,000}$$
$$= 25,000 + 1,923$$
$$= 26,923 \ \Omega$$

Therefore, the current flow through the circuit is changed to:

$$I = \frac{E}{R}$$
$$= \frac{9}{26,923}$$
$$= 0.00033 \ A$$
$$= 0.33 \ mA$$

So the voltage drop we read across resistor $R_2$ is:

$$E = IR$$
$$= 0.00033 \times 1923$$
$$= 0.54 \ V$$

That is ridiculously off from the expected 6 V we should get.

Conditions improve if a VOM with an input impedance of 50 k (50,000 Ω) is used.

$$R_t = 25,000 + \frac{50,000 \times 50,000}{50,000 + 50,000}$$
$$= 25,000 + \frac{2,500,000,000}{100,000}$$
$$= 25,000 + 25,000$$
$$= 50,000 \ \Omega$$

$$I = \frac{9}{50,000}$$
$$= 0.00018 \ A$$
$$= 0.18 \ mA$$

$$E = IR_2$$
$$= 0.00018 \times 25,000$$
$$= 4.5 \ V$$

better, but it is still hardly an accurate voltage reading.

Now, let's see what sort of reading we'll get with a DMM that has an input impedance of 1 MΩ:

$$R_t = 25,000 + \frac{50,000 \times 1,000,000}{50,000 + 1,000,000}$$
$$= 25,000 + \frac{50,000,000,000}{1,050,000}$$
$$= 25,000 + 47,619$$
$$= 72,619 \; \Omega$$

$$I = \frac{9}{72,619}$$
$$= 0.00012 \; A$$
$$= 0.12 \; mA$$

$$E = IR_2$$
$$= 0.00012 \times 47,619$$
$$= 5.4 \; V$$

This is reasonably close with the DMM, especially compared to our earlier examples. Actually, this is very much a worse-case scenario, and you usually won't run across such extreme errors in practical electronic work.

The input impedance of a multimeter is not fixed. Instead, it is so many ohms-per-volt. The input impedance is multiplied by the range setting. For instance, if you were using a 10 V setting on each of the multimeters in this example, the impedance of the 2 k VOM would be more like 20 k. Similarly, the 50 k VOM would have a practical input impedance of about 500 k on this range, and the 1 MΩ DMM would have an effective input impedance of about 10 MΩ.

Digital multimeters usually offer a number of extra functions not available on most analog multimeters. These features aren't essential, but they can be nice to have and don't add appreciably to the overall cost. A conductivity test function seems to be quite common on modern DMMs. *Conductivity* is simply the reciprocal of resistance;

$$\text{Conductivity} = \frac{1}{\text{Resistance}}$$

Conductivity is measured in units called *mhos*. Of course, mho is simply ohm backwards. Conductivity measurements are useful for very low resistances. For example, 22,500 mhos is a lot more convenient value than 0.0000444 Ω. In practical electronics work, however, the conductivity test function is of very limited practical value. It's certainly not a feature that most electronics technicians should want to pay extra for. Today, electronics technicians use the unit *Siemens*, instead of mhos.

DMMs also often include functions for measuring such electrical parameters as capacitance and frequency, as well as specialized diode and transistor testing. Until recently, digital multimeters were considerably more expensive than their analog equivalents, but now a good DMM can be purchased for under $50. I've seen some in the $20 to $30 range.

The analog multimeter is, however, far from obsolete. For some testing procedures, an "old-fashioned" VOM will do a lot better job than an up-to-date model DMM. For many electronic tests, the exact, multidigit precision of a DMM isn't really needed. For example, it probably doesn't matter too much if a circuit's power supply is putting out 9.12 V or 8.93 V, instead of exactly 9.00 V.

In many practical testing procedures, the exact measured value isn't as significant as the amount of change in the value over a short period of time. An analog VOM would certainly be a better choice than a DMM for such applications.

An example of this type of test would be the use of an ohmmeter to test a capacitor. (Refer to chapter 4.) Essentially, when the ohmmeter's test voltage is first applied across the capacitor's leads, the pointer should jump to a very low value, then slowly move back to a higher resistance value as the capacitor is charged. This process is very clearly visible on an analog meter, but on a DMM, the result would be just a blur of numbers changing too rapidly to be read. The test would be meaningless on a DMM.

Ideally, if you do more than casual work in electronics, you should own both a digital multimeter and an analog multimeter. Although there is considerable overlap in their functions, they are each often good for different purposes. If you can't afford both, we recommend that you go with an analog VOM or FET voltmeter first. The DMM can wait until you can afford it. Generally speaking, an analog multimeter will tend to be more versatile. If you are working seriously enough in electronics to require the precision of a digital readout, you definitely should invest in a simple analog VOM as a backup. It won't go to waste.

This is not to say that any electronics technician wouldn't find real advantages in having a DMM handy, but the advantages tend to be in the realm of luxuries, rather than absolute necessities. There is a tendency today, fostered by advertisers, that digital is automatically better for everything, and that isn't true.

Even if you primarily service digital circuitry, an analog VOM would still be the better choice if you had to restrict yourself to just a single multimeter. Fortunately, prices have come down sufficiently that few of us really have to make an either/or choice. It's not all that expensive to buy both an analog VOM and a DMM.

Digital circuitry is great, but it is important to keep things in perspective. There is no reason to throw out all analog circuitry and devices. In fact, there are often good reasons not to do so. Analog is better for some things than digital, just as digital is better for other purposes. It makes good sense to use both technologies, choosing the one most appropriate to the specific task at hand.

A well-stocked electronics workbench should have at least an analog VOM and a DMM. An analog VTVM or EMM would also be very desirable. If you can afford it, it is often useful to have several VOMs with spring-loaded clip leads. You can then monitor different parts of a circuit simultaneously.

# Adapters for multimeters

The standard multimeter measures volts, ohms, and milliamperes, so it is quite a versatile device. It can be made even more versatile with special add-on adapters to permit additional types of measurements. Usually these adapters convert some other parameter into a proportional voltage.

These adapters need not be particularly complex. For example, an ordinary semiconductor diode can be used as a simple temperature-to-voltage converter. The scale won't be linear over a very wide range, but in some applications, it is sufficient.

A number of multimeter adapters are available commercially, especially from manufacturers of DMMs. Usually, however, you have to build one yourself. Plans can be found for these adapters in the popular electronics magazines (*Radio Electronics, Modern Electronics, Hands-On Electronics,* etc.).

Typical multimeter adapters create a proportional dc voltage from such parameters as:

- temperature
- capacitance
- frequency
- light intensity

Other adapters extend the range of the multimeter, permitting you to read very large or very small voltages and/or currents. Adapters are also available to accurately measure ac voltages above the basic line frequency (60 Hz). For example, the circuit shown in Fig. 1-4 converts an ac voltage in the rf range into a proportionate dc voltage. Adapters for measuring true root-mean-square (rms) values by converting them to a proportionate dc voltage are discussed in chapter 8.

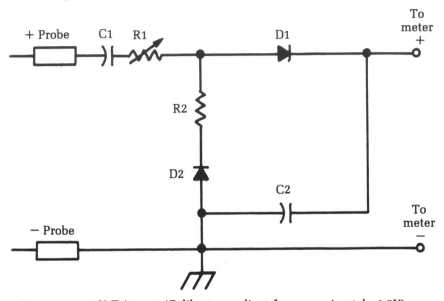

| R1 | 10K Trim pot (Calibrate—adjust for approximately 6.5K) |
| R2 | 47K Resistor |
| C1 | 0.82 $\mu$F Capacitor |
| C2 | 120 pF Capacitor |
| D1,D2 | Small signal diodes |

**Fig. 1-4**   *This circuit converts an ac voltage in the RF range to a proportionate dc voltage.*

# Oscilloscopes

Probably the most important and useful piece of electronic test equipment is the multimeter, but the oscilloscope runs a very close second. It is particularly helpful when dealing with signals that change over time. You can use an oscilloscope to measure dc voltages, but this method will usually be technological overkill. A multimeter (either digital or analog) almost certainly would be more convenient for the job. When it comes to ac signals, however, the oscilloscope wins hands down over the multimeter.

Most multimeters, if they can measure ac voltages at all, can give an accurate reading only if the signal in question is a relatively pure sine wave, with no harmonic content or other overtones. A more complex waveform will tend to confuse the meter because an ac voltage is not a straight-forward value like a dc voltage. For example, 3 volts dc (Vdc) is 3 volts dc, and that is that, but an ac voltage, by definition, changes values continuously over time in a repeating pattern. A 3.0 volt ac (Vac) signal will probably have instantaneous values like 2.3 V, 0 V, − 0.7 V, or even 4.1 V, depending on what point in the cycle you are measuring.

Sometimes an ac voltage is measured as a *peak value*, the maximum instantaneous voltage the waveform ever achieves during its cycle (Fig. 1-5). Notice there are two peak voltages: a positive peak voltage, and a negative peak voltage. Assuming the waveform is symmetrical around true zero (ground potential), these two peak voltages will be equal, except for their opposing polarities. The peak voltage of an ac waveform can be useful in determining whether the maximum ratings of sensitive components are ever exceeded, but it doesn't really tell much about the waveform signal itself. In addition, it would be very difficult to design a circuit that would display the peak voltage value on an analog meter or a digital read-out.

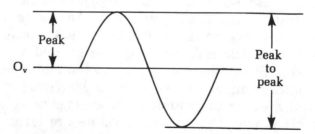

**Fig. 1-5**    *The peak voltage is the maximum instantaneous voltage occurring during the waveform's cycle.*

Often, an ac waveform will not be symmetrical around true zero. That is, a given waveform might have a positive peak voltage of 4.2 V and a negative peak voltage of − 1.8 V. Just saying the peak voltage of the waveform is 4.2 V would be quite misleading for most purposes. A peak to peak voltage will, at least, give more of an idea of the waveform's magnitude. The *peak to peak voltage* is simply equal to the difference between the positive and negative peak voltages:

$$E^{pp} = E_{p+} - E_{p-}$$

where $E_{pp}$ is the peak to peak voltage, $E_{p+}$ is *the positive peak volt-age, and* $E_{p-}$ is the negative peak voltage.

Four our example waveform with a positive peak voltage of 4.2 V and a negative peak voltage of −1.8 V, the peak to peak voltage is equal to:

$$E_{pp} = 4.2 - (-1.8)$$
$$= 4.2 + 1.8$$
$$= 6.0 \text{ V, peak to peak}$$

Unfortunately, the peak to peak voltage still doesn't tell very much about how the ac signal will function in a given circuit. Standard formulae, such as Ohm's law, will not work. Part of the problem is that, for most ac waveforms, the peak voltages occur for only a very brief portion of each cycle. (There are exceptions, of course. A square wave is at its positive peak voltage for half of each cycle, and at its negative peak voltage the rest of the time.)

To get a meaningful ac voltage value for the waveform as a whole, it is generally necessary to take some sort of average. Several instantaneous voltages per cycle are added together and the sum is divided by the number of samples. Unfortunately, for a symmetrical waveform, the complete average will always work out to 0 (plus or minus minor rounding off errors). Obviously, this value is utterly useless. An average voltage is therefore normally taken over just one half of each waveform. For a sine wave, the average voltage always works out to 0.636 times the peak voltage. Other waveforms will have different average values.

Even with the average voltage, Ohm's law and other important electronics formulae still do not work. Moreover, it requires some complicated circuitry to measure average ac voltages.

For most practical purposes, ac values are given in root-mean-square (rms). Such values are calculated through some rather complex mathematics. For sine waves (but no other waveforms) the rms value equals 1.11 times the average, or 0.707 times the peak. Peak, average, and rms values are compared in Table 1-1.

The big advantage of rms values is that they allow us to use Ohm's law and other common equations just as we would in a dc circuit. According to Ohm's law, the current flow is equal to the voltage divided by the resistance. A rms ac voltage will give the same current value as the comparable dc voltage. For example, let's say we have 10 Vdc being dropped across a 250 Ω resistance.

**Table 1-1   Ac voltages can be measured as peak, average, or rms values.**

| = | Peak | Peak to peak | Average | RMS |
|---|---|---|---|---|
| Peak | — | × 0.5 | × 1.57 | × 1.41 |
| Peak-to-peak | × 2 | — | × 3.14 | × 2.82 |
| Average | × 0.636 | × 0.318 | — | × 0.9 |
| RMS | × 0.707 | × 0.3535 | × 1.11 | — |

Find the known value unit in the top line. Follow the column down until it intersects the line for the desired value unit. Multiply by the constant in the space. For example, if you know the average value and want to find the rms value, multiply the average value by 1.11.

The dc current is therefore equal to:

$$I = \frac{E}{R}$$

$$= \frac{10}{250}$$

$$= 0.04 \text{ A}$$

If we replace the 10 Vdc with a sine wave of 10 V rms through the same 250 Ω resistance, we will also get an ac current value of 0.04 A rms. However, the average voltage for that same ac signal would be about 9 V, and the peak value would be 14.1 V, resulting in rather meaningless current values if we try to use Ohm's law.

It is relatively easy to design an ac voltmeter to measure the rms values of sine waves, but such a meter will not be able to give accurate readings for any other waveform, such as a triangle wave or a square wave. Even worse, the ac voltmeter will not give you any indication of whether the signal being monitored is reasonably close to a sine wave or not. It just assumes the input signal is in the form of a sine wave and measures it accordingly. Clearly, this method will give some totally meaningless readings for certain waveforms.

With an oscilloscope, you can see the actual waveform. You can easily see the peak points and its zero lines. You can see whether or not the waveform is symmetrical. If the ac waveform is riding on a dc voltage, this is also clearly visible. An oscilloscope permits you to actually see the signal being monitored, by drawing a picture of it on a CRT screen (similar to that of a television set.) The oscilloscope, therefore, permits us to perform many test procedures that would be impossible or impractical with a multimeter. Chapter 5 covers a number of common test procedures using the oscilloscope.

Like most other types of electronic equipment, oscilloscopes range from relatively simple to very complex devices. They are available with numerous specialized features, which might or might not be useful in the specific type of electronics work you do.

A simple, basic oscilloscope displays one waveform at a time, which is useful for many purposes, but sometimes it is not enough. You cannot compare phase relationships (timing of cycles) among various parts of the circuitry. It would also be difficult to determine whether a particular stage was adding subtle distortion (mishaping of the waveform) of an input signal. Under such conditions, a dual-trace oscilloscope is needed.

## Dual-trace oscilloscopes

A *dual-trace oscilloscope*, sometimes called a *dual oscilloscope*, is essentially two oscilloscopes in one. As the name suggests, two signal traces are displayed on the CRT screen, which is very useful for troubleshooting and servicing. Two signals can be compared directly. For example, you can monitor the input signal for a questionable stage and its output simultaneously, making it relatively easy to see if the stage in question is doing what it is supposed to do to the signal or if it is distorting or shifting the signal in some unintended way. In many electronic circuits, signal phase is crucial.

How can you find an out-of-phase signal with a single-trace oscilloscope? *Phase* is the description of the signal timing—when each cycle begins and ends—as compared to some reference. It is a comparative, rather than an absolute, parameter (such as voltage or resistance). A dual-trace oscilloscope permits you to make such comparisons.

The obvious way to design a dual-trace oscilloscope would be to build a CRT with two independent electron guns, each with its own set of deflector plates—in other words, two CRT tubes in a single glass envelope. Unfortunately, this approach can be very expensive, although it is used in some high-end dual oscilloscopes. Two CRT tubes also introduces a number of tricky design problems, such as how to keep the beam from one electron gun from being affected by the deflector plates of the other gun. It can be done, but it is not easy or inexpensive.

Most practical dual-trace oscilloscopes use some type of digital switching to simulate the effect of two electron beams, even though only a standard single-gun CRT is used. There are two basic ways this switching is done: alternate mode and chopped

mode. Most dual-trace oscilloscopes can operate in either switching mode, depending on the sensed signals and/or the front panel
control settings. A simplified block diagram of a dual-trace oscilloscope is shown in Fig. 1-6. The difference between the two
modes is in how the electronic switch is operated.

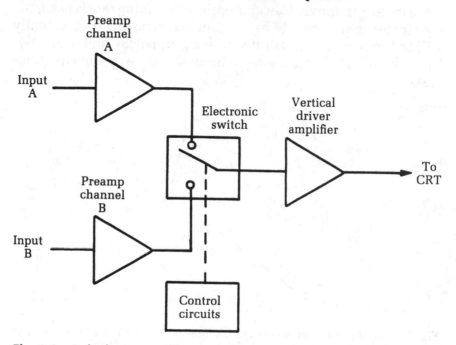

**Fig. 1-6**  *A dual-trace oscilloscope switches back and forth between two
input signals.*

In the alternate mode, trace A is first drawn for one complete
sweep cycle. That is, the beam is moved across the screen, just as
in a regular, single-trace oscilloscope, but at the end of the sweep
cycle, when the beam reaches the far end of the screen, input A is
switched out and input B is switched in. An offset voltage moves
the center (zero) line of this second trace, so the two traces do not
overlap (unless you choose to set the controls for such overlap, for
certain test functions). Now, trace B is drawn across the screen for
one complete sweep cycle. When the beam reaches the far end of
the screen, the input is switched back to signal A, and this trace is
redrawn for the next sweep cycle. The oscilloscope's CRT screen
is coated with a phosphorous with some delay. This coating, cou-
pled with persistence of vision, makes it appear that both traces
are being continuously displayed, unless the sweep frequency is
very low.

The chopped mode is rather similar, except that the switching takes place much more frequently. It does not wait until the end of a sweep cycle; instead, the beam is switched back and forth between the two traces many times during each sweep cycle. A little bit of trace A is drawn, then signal A is turned off and a little bit of trace B is drawn. If an appropriate switching rate is used, the two traces will appear to be continuous lines, but they actually will be broken up, as illustrated in exaggerated fashion in Fig. 1-7. These breaks in the traces might become visible at some frequencies.

**Fig. 1-7**    *In the chopped mode, the oscilloscope switches between the two signals many times per sweep cycle.*

For some signal pairs, the alternate mode will work better and provide more precise information about the signals. In other cases, the chopped mode is the better choice. Actually, for most noncritical work, the choice of mode probably won't matter too much — either will work fine.

Dual-trace scopes are now the norm, except for small portable scopes and very inexpensive bench scopes. Single-trace scopes are still manufactured and used, but they are becoming increasingly rare. In the future, there will be even more extensive multiple-image scopes.

## Storage oscilloscopes

Occasionally, while working on an electronic circuit, a technician might want to study certain displayed waveforms for awhile. With a continuous, unvarying periodic (ac) waveform, this is no problem, since the pattern is repeatedly redrawn during each cycle of the oscilloscope's sweep signal. One-shot signals, how-ever, such as those used in switching circuits, come and go quickly. Often the entire signal of interest lasts only for a few mil-liseconds, and then it's gone, leaving just a flat trace on the oscil-loscope's screen.

The persistence of the phosphors coating the inner surface of the oscilloscope's CRT screen permit them to glow long enough for the trace to be seen, but the displayed trace fades away within seconds. This often might not be long enough for the technician to read enough information from the displayed waveform. Ob-viously, this type of problem becomes more crucial as the signal increases in complexity and detail.

The solution to such problems is to use a special type of oscil-loscope known as the *storage oscilloscope*. A storage oscilloscope can be used as a regular oscilloscope, duplicating all the standard functions, but it also features a special storage mode to permit rel-atively long-term display of nonperiodic (not repeating) signals. To accomplish this long-term display, the storage oscilloscope uses a special type of CRT. The basic structure of this device is il-lustrated, in somewhat simplified form, in Fig. 1-8.

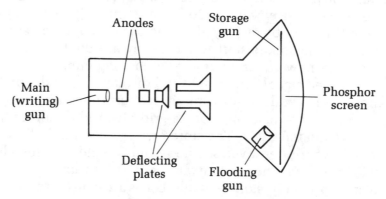

**Fig. 1-8**  *A storage oscilloscope uses a special type of CRT with two electron guns and a storage grid just inside the display screen.*

This type of storage CRT and the simpler CRT used in most basic oscilloscopes differ in two main respects. In the storage oscilloscope, the CRT has two electron guns, instead of just the usual one, and a storage grid of tiny electrodes mounted just in-side the phosphor-coated display screen. The two electron guns in the storage oscilloscope's CRT are called the *main gun* and the *flooding gun*.

The main gun functions just like the electron gun in an ordinary CRT. It produces the electron beam that draws the actual signal trace on the display screen. In the basic (nonstorage) mode, the flooding gun and the storage grid are left unused.

When the oscilloscope's storage mode is activated, the electrodes in the storage grid are saturated with a negative electrical charge just before the momentary signal trace is drawn on the display screen by the main gun. The electrodes that are struck by the directed electron beam from the main gun as it draws the trace take on a localized positive charge. The other electrodes in the storage grid, which are not touched by the main gun's electron beam, do not pick up a positive charge. When the complete desired waveform appears on the face of the CRT, the saturating negative charge is removed from the storage grid. The storage grid is specifically designed so that the positive charge on individual electrodes decays at a very slow rate, unless it is deliberately removed by resaturating the entire grid with an external negative charge.

The flooding gun now takes over, irradiating the entire storage grid, which prevents the electrons from the flooding gun from reaching the phosphor-coated screen, except in those areas holding a localized positive charge. As a result, the original waveform trace continues to be displayed on the oscilloscope's screen. As the glowing phosphors start to fade, they are continuously reenergized by new electrons from the flooding gun.

The stored signal can be held and displayed, usually up to a maximum of several minutes. Infinite storage times are not possible, of course, because the storage grid electrodes cannot hold their charge indefinitely. The stored charge inevitably leaks off within a few minutes. Even so, the stored waveform is held long enough for the technician to examine it quite thoroughly.

Not surprisingly, storage oscilloscopes are considerably more expensive than standard oscilloscopes. If you need the storage function, however, the increase in price is well worth it.

## Digital oscilloscopes

There is no question that the big trend in electronics over the last decade or so has been the general move from analog to digital circuitry. It should not be at all surprising that a number of digital oscilloscopes are now available.

Unlike the digital multimeters discussed earlier, the final output of a digital oscilloscope is still in analog form — the waveshape is visually drawn, not represented by displayed numerical values. The difference is in how this is done. A digital oscilloscope uses digital circuitry to prepare the signal for display. In a digital oscilloscope, the analog signal to be monitored is converted into digital form for processing and multiplexing, then it is converted back into an analog image for the final display.

Some digital oscilloscopes do not even have a CRT. A multiplexed arrangement of light-emitting diodes (LEDs) often is used to display the waveform. By lighting up only the appropriate LEDs in the matrix, it is possible to draw the waveform in dot-to-dot fashion. Obviously, the more LEDs there are in the matrix, the clearer the image will be and the more detail will be possible.

More recently, liquid crystal displays (LCDs) have been used for this purpose, permitting lower power consumption, simpler multiplexing circuitry, and less expensive construction (each individual LED has to be soldered into the circuit). The display signal data must be in digital form to light up the appropriate LED or activate the appropriate spot on a LCD in a display matrix.

An LED display tends to offer a little less detail than a CRT, but it is less bulky and less fragile. It also does not require the high operating voltages a standard CRT demands. A LCD display can be even smaller and consume even less power than an LED display. Hand-held dual-trace oscilloscopes are available with LCD displays. A hand-held unit would not be practical using a standard CRT.

Unlike analog and digital multimeters, there would be no particular advantage for most technicians to own both an analog oscilloscope and a digital oscilloscope. They both perform essentially the same functions, just using different circuitry.

A special type of digital oscilloscope may use either a CRT or an LED/LCD display, but the digital circuitry is employed for signal processing purposes. The digital storage oscilloscope (DSO) is becoming increasingly popular among serious electronics technicians; however, the cost and expense of such instruments place them outside the reach of most electronics hobbyists.

The DSO takes the idea of the analog storage oscilloscope one step further. In an analog storage oscilloscope, the stored signal will fade away after a few minutes. A DSO, on the other hand, can hold the displayed signal indefinitely, as long as power is continuously supplied to the instrument. The data about the displayed signal is stored in numerical form in a digital memory, not just in electrical charges within the display screen. If a DSO is combined with a computer, as is often done, the signal can even be recorded for permanent storage and later recall and reuse.

Because the stored signal is in digital form, it can be manipulated in a number of ways. For example, certain features of the displayed waveform can be highlighted for emphasis, or the technician can enlarge the display to "zoom in" on areas of particular interest. The stored signal can be combined or compared with some other reference signal stored at an earlier time. Another advantage of the DSO is that, as a rule, digital timing can be more accurate than analog timing.

When shopping for a digital storage oscilloscope, all other factors being equal, choose the DSO with the longest record length specification. The record length determines how long the stored waveform can be; that is, how much information the DSO can hold at a time.

In addition to the analog bandwidth specification (which is essentially the same for both digital and analog oscilloscopes), you will also need to consider the sampling rate. The analog input signals are converted into digital form by sampling their instantaneous voltages many times per second. The speed at which this is done is called the *sampling rate*, and the faster the better. The sampling rate is usually given in the form of the highest usable signal frequency for the DSO's maximum sampling rate.

According to the Nyquist theorem, the sampling rate must be at least twice the frequency of the signal to be digitized, or a form of distortion known as *aliasing* will occur. This distortion disguises the signal frequency and gives totally meaningless results.

If the sampling rate is just twice the signal frequency, the signal's frequency can be measured accurately, but all details of the actual waveshape will be lost. For this reason, some manufacturers of DSOs specify the sampling rate for four or more points per cycle or period. Before comparing sampling rate specifications for competing DSOs, make sure you know how the figure was derived. The four-points system will give less impressive numbers than the straight Nyquist method, but in practical terms, the four-points system will offer more accurate results.

Several different sampling methods are used in DSOs. Most currently available DSOs offer the options of real-time sampling and repetitive sampling. Real-time sampling uses a very high sampling rate to provide enough sample points in a single sweep to permit an accurate reproduction of the waveform. This type of sampling is used primarily for one-shot or aperiodic (nonrepeating) events, such as switching signals.

For periodic (ac) waveforms, repetitive sampling is generally preferred. Because the monitored waveform keeps repeating itself, fewer samples can be taken during each individual sweep, and the waveform can be reconstructed in detail by combining the sample points for several consecutive sweeps. A lower actual sampling rate can be used to give the same degree of display detail.

Practical test procedures using oscilloscopes of various types are covered in chapter 5.

# Signal generators

A *signal generator* is used to provide a known input signal for signal-tracing type tests. The standard procedure for signal tracing is to feed the signal generator's output to the input of the circuit to be tested. In some cases, it is necessary to disconnect the circuit's normal input connection. A measurement is then taken at the output of the first stage of the circuit.

Depending on what is being tested, look to see if the signal is missing, overly attenuated, distorted, or clipped. If the signal looks okay at the output of the first stage, move on to the output of the second stage. Continue with this process, moving toward the circuit's final output. When you find a point where the signal disappears or otherwise becomes unacceptable, you have isolated a trouble stage. Look for a faulty component in that stage of the circuit. All earlier stages have been cleared of suspicion.

Most signal generators put out a single-frequency signal of a specific waveshape (sine waves, square waves, and triangular waves are the most commonly used). Some signal generators sweep through a range of frequencies and used to test the frequency response of the circuit under test.

There are two broad classes of signal generators: AF signal generators and RF signal generators. The chief difference between the two signal generators is in the frequency of the generated signal. An AF signal generator produces a fairly simple, continuous waveform nominally in the audio frequency range. The theoretical range of audible frequencies runs from about 20 Hz to 20 kHz (20,000 Hz); however, many AF signal generators can generate signal frequencies below 20 Hz, and often well above 20 kHz. Modern technology permits the design of a single circuit with a range from close to 0 Hz (dc) up to 100 kHz (100,000 Hz) or so. This is still considered to be an AF signal generator because the basic design is that of an AF circuit.

An RF, or radio frequency, signal generator, on the other hand, produces signals with much higher frequencies, sometimes ranging well into the megahertz (millions of Hz) range. Such high frequencies require special design techniques in the circuitry. Another important difference is that an RF signal generator usually has some sort of simple modulation—that is, the main RF signal will be combined with a simple AF signal (usually internally generated) to simulate a radio broadcast.

## AF signal generators

Within the broad category of AF signal generator are three basic types of instruments, although the use of the terminology sometimes overlaps. An AF signal generator will be an audio oscillator, a signal generator, or a function generator. There are real differences among these categories, even though, in practice the names are sometimes used interchangeably. In some cases, the dividing line between these three classes of AF signal generators is not hard and fast.

**Audio oscillators**   Generally speaking, an *audio oscillator* produces an AF sine wave. A sine wave, shown in Fig. 1-9, is the simplest and purest ac waveform. Unlike all other waveforms, the sine wave consists of just a single frequency component, with no harmonic content at all. It is relatively difficult to generate a good, pure sine wave with no distortion (unwanted harmonic content), so audio oscillators are usually fairly expensive and sophisticated instruments.

Because a sine wave is so pure, it is helpful in identifying harmonic distortion and noise generation in the circuit under test. The output from the audio oscillator is used as the stage's input signal, instead of its normal signal source. This way, you know exactly what the signal going into the circuit is like, and you can compare the results at the output.

**Fig. 1-9**   *An audio oscillator gen erates an audio frequency sine wave.*

**Fig. 1-10**   *Another common electronic waveform is the triangle wave.*

**Signal generators**   The term *signal generator* is usually employed for an instrument that produces some waveform other than a sine wave. Typical examples are the triangle wave (Fig. 1-10), the sawtooth wave (Fig. 1-11), and the rectangle wave (Fig. 1-12).

**Fig. 1-11**   *A sawtooth wave.*     **Fig. 1-12**   *A typical rectangle wave.*

These different waveforms are useful for different types of tests. A triangle wave is often used in place of a true audio oscillator (sine wave generator) because the required circuitry tends to be simpler and less expensive. Also, the harmonic content of a triangle wave is relatively weak.

All ac waveforms except the sine wave consist of multiple frequency components. The base frequency is the *fundamental frequency*, the frequency of the waveform as a whole. It is the strongest (highest amplitude) frequency component. In simple waveforms, like the ones we are concerned with, the fundamental is also the lowest frequency component.

The other frequency components are called *overtones*, or *harmonics*. A harmonic is an exact integer multiple of the fundamental frequency. (Some complex waveforms have enharmonic overtones, which are higher than the fundamental frequency but are not exact integer multiples.) The harmonics are numbered by their integer multiples. For example, if the fundamental frequency is 250 Hz, then the following harmonics are possible:

| Harmonics | Hertz |
|:---:|:---:|
| 2nd | 500 |
| 3rd | 750 |
| 4th | 1000 |
| 5th | 1250 |
| 6th | 1500 |
| 7th | 1750 |
| 8th | 2000 |
| 9th | 2250 |
| 10th | 2500 |

and so forth, continuing indefinitely.

The harmonics are weaker (lower amplitude) than the fundamental frequency. The higher the harmonic, the weaker its amplitude. The highest harmonics are generally so weak that their effect on the overall waveform is negligible and they can reasonably be ignored. The specific harmonics included and their relative amplitudes determine the waveform.

For example, a triangle wave includes the fundamental and all odd harmonics, but no even harmonics. A square wave includes the same harmonics as a triangle wave of the same frequency, but the amplitude of each individual harmonic is much stronger in the square wave. A sawtooth wave includes all of the available harmonics, both odd and even.

A signal generator that produces rectangle waves is often called a *pulse generator*. (The terms *rectangle wave* and *pulse wave* are generally synonymous.) A rectangle wave has just two voltage levels, HIGH and LOW, with no transition time between the two. (Most other waveforms slide linearly through a range of voltages.)

Most modern pulse generators feature a control to adjust the duty cycle of the rectangle wave. The duty cycle determines how much of each complete cycle will be in the HIGH state, thus affecting the harmonic content of the waveform. The duty cycle is usually expressed as a ratio, 1:$X$. Every harmonic that is a multiple of $X$ is omitted from the waveform. For example, if the duty cycle is 1:3, every third harmonic will be omitted: Fundamental, 2nd harmonic, 4th harmonic, 5th harmonic, 7th harmonic, 8th harmonic, 10th harmonic, and so forth. Similarly, if the duty cycle is 1:4, any harmonic that is a whole number multiple of 4 will be left out of the sequence: Fundamental, 2nd harmonic, 3rd harmonic, 5th harmonic, 6th harmonic, 7th harmonic, 9th harmonic, and so forth.

The popular and widely used square wave is just a special form of the rectangle wave, in which the level is HIGH for exactly half of each complete cycle. The duty cycle is 1:2; therefore, all harmonics that are multiples of two (all even harmonics) are omitted.

Sometimes you will see a reference to a test instrument known as a *signal injector*. This is just a simple AF signal generator that feeds its output through a probe that can be conveniently moved through the circuit under test to check out various points. Usually a signal generator will have very few controls and features.

**Function generators**    The third major type of AF signal generator is the *function generator*. A function generator is basically a special form of the signal generator. Most signal generators put out just a single waveshape. A function generator, on the other hand, lets the technician choose from three or four different waveshapes. For certain tests, a square wave, for example, might be better than a sine wave, or vice versa.

Function generators also tend to have a wider frequency range than simple signal generators. Many function generators are capable of putting out signals ranging from a fraction of a hertz up to several megahertz.

### RF signal generators

For radio frequency work, an RF signal generator is used. It is similar in concept to regular (audio and ultra-audio) signal generators, but the output frequency is much higher. The output from an RF signal generator is almost always a sine wave. Usually there is the capability to modulate the main output signal with another signal in the audio range, and the technician can select between amplitude modulation (AM) or frequency modulation (FM).

Because of the high frequencies being produced, an RF signal generator must be very well shielded. Usually it must be type-approved by the FCC. RF signal generators are usually crystal-controlled for accuracy and stability.

## Signal tracers

A *signal tracer* is a device for detecting the presence of a signal. Usually it's not much more than a small amplifier and speaker with test leads for taking its input from the circuit point under test. A signal tracer is often used with a signal generator to test audio equipment, but it can be used alone, also.

A signal tracer is often used in the reverse manner for signal generators. The signal tracer is first connected to the circuit's output (usually the speaker terminals). It is progressively moved backward through the circuit, stage by stage, until the signal has been found. If there is no signal at the output of stage 4, but the signal is fine at the output of stage 3, then the problem is in stage 4.

## Capacitance meters

Until fairly recently, capacitor testers were not found on too many electronics servicing workbenches for several reasons. The early capacitor testers were bulky and expensive. More importantly, they really weren't good for very much. Most were basically "go/no-go" type testers that checked for shorts or opens and perhaps gave a rough measurement of leakage. Some of the top units also gave a (very) approximate reading of the capacitance range.

Shorted and open capacitors almost always can be pinpointed in-circuit without a capacitor tester. It also isn't very difficult to isolate a leaky capacitor with in-circuit voltage and current tests. The early capacitor testers could be handy in certain circumstances, but for most technicians they were an unnecessary luxury. It was nice to have one, but it was no hardship to live without it.

The modern capacitance meter, however, is completely different. It is essentially a digital device that displays the actual capacitance numerically and with great precision. Most capacitance meters work by measuring the time constant of the capacitor being tested. The *time constant* is the time required for the capacitor to charge up to two-thirds of the applied voltage through a given resistance. The resistance, of course, is determined by the meter's circuitry and is a constant.

For many applications, the exact value of capacitance is not terribly important. Circuit resistances are usually much more crucial. Most resistors used in circuits today have a tolerance of 5 percent, while 20 percent tolerances are typical for capacitors. However, sometimes a capacitor can change its value outside its tolerance range and throw off circuit performance.

The capacitance value does become crucial in frequency-determining circuits, such as filters, tuners, and oscillators. In some applications, the actual capacitances don't matter as much as whether or not two or more capacitances in the circuit are closely matched.

A capacitance meter is also useful for hunting down stray capacitances in a circuit. Remember, a *capacitor* is really nothing more than two conductors separated by a dielectric (insulator). Stray capacitances can show up almost anywhere. Even air can serve as a dielectric. Sometimes stray capacitances are at fault in older equipment that has worked fine for years but suddenly starts behavior erratically. When the stray capacitance is located, it can usually be eliminated by repositioning connecting wires or replacing or adding shielding between the two conductors that are acting like capacitor plates. Adjacent traces on a pc board are particularly prone to stray capacitances.

Another use for a capacitance meter is to test the quality of a length of cable. Most standard types of cable (coax, twin-lead, ribbon, etc.) have characteristic capacitances per foot (or per meter). Just measure the capacitance across the cable and divide by the length to get the specified capacitance-per-unit-length value, or something very close to it.

$$\text{Capacitance-per-unit-length} = \frac{\text{Total Capacitance}}{\text{Length (in number of units}}$$

Make sure that you use the same length unit for both halves of the equation.

This equation can also be reversed to get an estimate of an unknown length of presumably good cable.

$$\text{Length (in number of units)} = \frac{\text{Total Capacitance}}{\text{Cap.-per-unit-length}}$$

As you can see, the modern capacitance meter is a truly versatile and useful piece of test equipment, made possible by digital technology.

# Frequency counters

It seems the frequency counter is the status symbol for electronics technicians. Everybody wants one, even if they're not entirely sure what they want if for. If you work primarily in general TV servicing, you probably wouldn't use a frequency counter very often. On the rare occasions when you need to measure a frequency, you can use your oscilloscope. Just count the number of cycles displayed and divide by the time base.

On the other hand, in some applications a frequency counter can be extremely valuable. Two such applications include musical instrument servicing and working with RF transmitters. In RF transmitters, precise frequencies are required by law. A frequency counter is a handy tool for making sure that oscillators, multipliers, and frequency synthesizers are working properly and are tuned correctly.

All modern frequency counters are digital devices. The measured frequency is displayed directly in numerical form. The typical input impedance of a frequency counter is about 1 megohm (M$\Omega$), so loading is minimal. Most commercially available frequency counters measure frequencies up to about 50 MHz or 100 MHz, which should be sufficient for most servicing work. However, if you need to measure higher frequencies, accessory prescalers are available to extend the range of a frequency counter.

# Servicing test equipment

Any piece of electronic equipment is likely to require servicing sooner or later. Though most people don't think about it, that obviously includes test equipment.

Conceptually, at least, servicing a piece of test equipment is no more difficult than any other type of electronic circuit. Two special types of difficulties are involved. The first is that sometimes you need the instrument being tested to perform the necessary tests. Whenever possible, it is a good idea to have spares of general-purpose equipment, such as multimeters. You need a second multimeter to find a defective component in your first multimeter. In addition, if you have only one multimeter and it breaks, virtually all of your service work will come to a standstill until you get it fixed.

Multimeters are inexpensive enough that most technicians can afford two or three. The spare units need not be deluxe models with all of the special features of your primary meter. They just need to be sufficient to get you by in an emergency.

Some types of equipment are too expensive to have spares on hand. For example, unless you work in a large shop with several staff technicians, you probably will have just one oscilloscope. If it develops a fault, you will need to borrow or rent one until yours can be repaired or replaced. It is a good idea to check out the availability of an emergency replacement before you are faced with an actual emergency.

VOMs are probably the easiest piece of test equipment to repair because there really isn't very much inside them. They are basically made up of switches, a handful of precision resistors, a diode or two, and indicators. Common faults include broken wires leading from the switches, shorts, or burned-out diodes or resistors. Burned-out components are almost always caused by accidentally feeding too high a voltage or current to the VOM or trying to operate it on the wrong range. Trying to measure a moderate to large ac voltage while the meter is set for ohms, for example, almost surely will result in disaster. Unfortunately, this is very easy to do. Almost every experienced technician I've ever met has blown out at least one meter at some point in his career.

Unless the over-voltage or current is extremely high, this type of problem is usually fairly easy to repair. It's just a matter of locating the burned-out component(s) and replacing them. Since there aren't that many components in the entire VOM circuit, you could test every component individually in just an hour or so. You can make better use of your time, however, by using a little logic to eliminate some of the components as suspects.

In most cases, the damaged VOM will work on some ranges or functions but not on others. Obviously any component included in a working range must be good, so there is no point checking it. Often one or two resistors or diodes will be visibly burned. Sometimes a burned component looks okay, but can be located by smell, especially right after it was damaged. But don't rely on eyeball or nose tests alone. Multiple components might have been damaged. A resistor might change value or be internally cracked without being visibly burned. Bad diodes usually can't be detected without a resistance test of the pn junction. A component that looks bad probably is, but a component that looks good might not be. Don't be too quick to jump to conclusions.

There is one part of the VOM that is tough to repair: the meter itself. It can be damaged by too large a signal or by a physical shock (such as being dropped). Occasionally, the only problem is a bent pointer needle. If you have a very sure hand, you might try repairing it. Open up the meter's housing and carefully bend the meter back into the correct position. Do not use too much force, or you could break off the pointer needle or bend it irreparably. The pointer needle must be absolutely straight, or the meter will be useless.

In some damaged meters, the driving coil might be burned out. It is usually not repairable, and the meter itself must be replaced. For a VOM, this usually means buying a whole new instrument, since the meter makes up the bulk of the unit's cost. Do not try substituting a cheaper meter unless you are really desperate. An unauthorized replacement probably won't fit mechanically very well, so it will be more prone to physical damage, and probably won't last very long. More importantly, its electrical characteristics probably won't be exactly the same as the original, adversely affect the accuracy of the unit.

VTVMs or other electronic multimeters (FETVMs, etc.) aren't much more complicated to service, but the circuitry does include one or more active amplification stages. These circuits can be repaired with standard voltage/current measurements and signal-tracing techniques. For some electronic multimeters, it might be worthwhile to order a replacement meter movement from the manufacturer, rather than junking the instrument and buying a new one. Use an exact replacement, or you'll just be asking for trouble.

Repairing a multimeter isn't particularly difficult in most cases. In order to put it back into service, however, you must first confirm its accuracy. The unit should be recalibrated after any repair or adjustment. It is a good idea to perform calibration procedures any time the instrument's case has been opened. Occasionally, the only repair needed on an apparently defective multimeter is recalibration.

First make sure the meter's pointer swings smoothly over its entire range. To check this, set the multimeter to one of its ohms ranges. With the test leads held apart, the pointer should be all the way over to the right end of the scale (infinity). Now touch the two lead probes together to create a short. The pointer should swing smoothly to the far left (zero). Bring the leads together and apart several times while watching the pointer closely. It should swing back and forth smoothly without sticking anywhere along its path. If it does stick, the most likely cause is some debris in the pivot base. Open the meter's housing and carefully remove any foreign matter. You'll need a sharp eye; any very tiny particle could cause a problem.

Now set the multimeter to a voltage setting and make certain it is reading the correct value. You need some sort of reference voltage—a battery will do, if it is fresh. A mercury cell is the best because it holds a very stable voltage, but an ordinary flashlight cell will do if it has never been used or hasn't been sitting on the shelf too long. A new flashlight battery puts out a reasonably precise 1.56 volts. Adjust the meter's calibration control (usually an internal trim pot) until the meter displays the correct reading.

An oscilloscope is best calibrated with a square-wave signal of a known level fed to the vertical amplifier inputs. Watch the display closely for any distortion, especially rounded corners. This could be caused by faulty coupling capacitors in the signal path.

Signal generators can be tested with an oscilloscope and/or a frequency counter. Make sure the output frequency correctly corresponds to the nominal frequency of the unit or the setting of the front panel control(s) on a variable unit. On an oscilloscope that is known to be correctly calibrated, check the output waveform for purity, low distortion, and symmetry.

# 2
# Dc voltage tests and power supplies

Dc VOLTAGE TESTS ARE USUALLY THE FIRST STEP IN TROUBLESHOOTing, since the dc voltages of a device usually provide the basis for operation. The fact that they are one of the easiest tests adds to this reasoning. The remainder of this chapter focuses on the wide range of power supplies and their circuits.

## Dc voltage measurements

Dc voltage is measured with a dc voltmeter. This sounds a little obvious, but there are several different types of voltmeters, and some circuits need the right type if test readings are to match a standard.

Three basic types of voltmeters are in wide use today. The first is the *analog voltmeter*, which uses a swinging needle over a calibrated scale (*d'Arsonval* movement). These meters have an adjustable resistance in series with the leads to allow proper range selection. The second major type is the *vacuum-tube voltmeter* (VTVM), or its transistorized counterpart, which uses an amplifier to drive the meter movement. The third major type is a relatively recent development. The *digital voltmeter* uses an analog-to-digital (A/D) converter to directly display the measured value in numeric form on an LED or LCD read-out panel. No mechanical meter movement is used.

In the analog type, the meter movement and its series resistance are hooked directly across the circuit under test. For measurements with a VTVM, a very large voltage-divider resistor is hooked across the circuit; the meter is driven by a dc amplifier whose input is "tapped down" on this voltage divider to obtain the desired range. Digital voltmeters are generally used in the same manner as mechanical VOMs. That is, the test leads are connected directly across the circuit or component to be tested. Each type of meter has its own uses, advantages, and disadvantages.

When it comes to using voltmeters, the most importance difference between the VOM (analog) type and the VTVM (tube) type is in their *input impedance*, or resistance. The input impedance of the meter sometimes can affect the voltage reading you get. The 1,000 $\Omega$ per-volt meter — a typical analog voltmeter — must offer a total resistance of 1,000 $\Omega$ for every volt of the full-scale range you want to be able to read. Such a meter, set to a 10 V (full-scale) range, would have a total resistance of only 10,000 $\Omega$; on a 200 V range, it would have a resistance of 200,000 $\Omega$. the 20,000 $\Omega$ per-volt meter — another analog voltmeter — is much more sensitive. With this meter, the 200 V range has a total resistance of 4 million $\Omega$ (4 M$\Omega$).

The unofficial standard for VOMs is 20,000 $\Omega$ per volt. Many modern VOMs have input resistance ratings of 50,000 $\Omega$ per volt, or even 100,000 $\Omega$ per volt. The earlier 1,000 $\Omega$ per volt VOMs are rarely used, except in very noncritical applications.

# Problems with circuit breakers

A circuit breaker is essentially a resettable fuse. When it blows, it doesn't have to be replaced. A manual push button resets the device. In most cases, the circuit breaker was tripped by a temporary transient on the ac power lines. If, once reset, everything seems to be working just fine, there is probably nothing to worry about. You don't know precisely what caused the problem (what made the circuit breaker trip), but whatever it was, it is gone now. It was presumably a transient phenomenon.

In some cases, however, a circuit breaker will be tripped each time it is reset, sometimes immediately every time power is applied to the protected circuit. In other cases, the equipment will work for a brief time, then the circuit breaker will mysteriously be tripped again.

If a circuit breaker repeatedly trips, or a fuse repeatedly blows, there is something wrong. Usually something within the circuitry is drawing too much current. There might not be an actual short circuit, but something is trying to consume too much power, resulting in the overload.

Surprisingly, this is not always the case. Sometimes circuit breakers go bad themselves and start misbehaving. How can you tell if you've got a bad circuit breaker or not?

When the questionable circuit breaker trips, clip a second circuit breaker across the first one's terminals. In other words, you now have two circuit breakers in parallel. One is the original one, which has tripped. The second is a circuit breaker known to be good and is reset (not tripped). Of course, this replacement circuit breaker must have the exact specifications as the original unit. If the new, parallel circuit breaker trips, too, then there is clearly some sort of current overload in the circuitry. The original circuit breaker has been operating properly. If the second, parallel circuit breaker does not trip, however, and lets the equipment operate normally, then the original circuit breaker must be defective. Simply replace it with the new unit you used for the test procedure.

# Checking for a hot chassis

In some electronic equipment, especially older devices, a metal chassis or case is used, and it is electrically grounded. For battery-operated equipment a metal case is rarely of any particular significance. If ac power is used, however, the common wire from the ac plug might be connected directly to the chassis. But if the connections are reversed, usually by inserting the plug upside down, the result can be a very dangerous shock hazard.

Because of the potential danger, this type of construction is seldom, if ever, used in modern electronics. When a distinction must be made between a common wire and a live wire, a polarized plug will be used, but they are a relatively recent development. Many consumers are quite ingenious in finding ways to invert a polarized plug or defeat the extra grounding offered by a three-prong plug.

You might occasionally find you have to work with an older piece of equipment with a grounded chassis. Even in modern equipment, physical damage might result in a short circuit to a metallic case or control panel. Obviously, a smart electronic technician will want to take precautions. If you touch a hot chassis while your body is grounded (for example, by touching a water pipe), you will receive a very painful, probably dangerous, and possibly even fatal electrical shock.

If an unpolarized plug is used, as on virtually all older electrical devices, there is a 50-50 chance that the wiring will be reversed, resulting in a live chassis. Even if there is a polarized plug, there is the possibility of an unintended short. Also, you have no guarantee the polarized plug was installed properly. We once serviced an old television where it turned out a customer had replaced a worn power cord. He used a cord with a polarized plug, but he installed it backward. This was a primary cause of the fault we were asked to service.

It is easy enough to test for a hot chassis, so there is no good reason not to. Just set your multimeter to measure ac voltage, with a range that will comfortably cover 120 Vac. (Usually the 150 V range will be the closest.) Make sure no power is applied to the system. Then, put on a pair of electrician's rubber gloves and individually connect one test lead to the chassis and the other to a good earth ground, such as a cold water pipe.

Once the test leads are connected, apply power to the questionable device and watch the meter. Ideally, nothing should happen. If you get a measurable ac voltage (probably equal to the line voltage), then you've got a hot chassis. Watch out!

Reverse the plug. If you now measure no voltage from the chassis to earth ground, you've identified the problem. It would be foolish to not replace the existing unpolarized plug with a suitable polarized plug.

If you get a voltage reading with either plug position or there is a polarized plug, especially if the measured voltage is considerably lower than the line voltage, then there is a mechanical short somewhere in the equipment being tested. Disconnect all power to the equipment and open it up. It should be fairly easy to find the short. Something conductive probably will be in physical contact with both the circuitry and the chassis. When you think you've found and corrected the short, repeat the hot chassis test as before. Sometimes you can be fooled. There might be a second short, or what you thought was surely the culprit wasn't causing the problem after all. Don't make assumptions and take foolish chances.

# All-polarity voltmeters

Voltmeters are quite easy to find, especially dc voltmeters, but ac voltmeters are hardly uncommon. On a VOM, DMM, or VTVM, a selector switch must be turned to the correct position (ac or dc). In some cases, one or both of the test leads must be plugged into different jacks on the test instrument. Of course, to measure a dc voltage, the meter's test leads must be arranged with the correct polarity, or the meter's pointer will be forced backwards, and be likely to be bent, or damaged.

On some multimeters, attempting to measure an ac voltage on a dc scale might result in damage to the meter's circuitry, especially if it is set for a too-low voltage scale. For most practical electronics work, this isn't really much of a hardship, although it might occasionally be a bit inconvenient. You simply have to pay attention to what you're doing ( a good idea under any circumstances), and be careful to hook up your voltmeter correctly.

Suppose, however, that you're not sure if the voltage you need to measure is ac or dc, or if it is dc, what the correct polarity is. You could try your luck, with the voltmeter set for its highest available rating first, but there will always be some inherent risk in such blind testing. Also, if you are performing a number of tests throughout a circuit in quick succession, it can be a major nuisance switching back and forth between the ac and dc settings and keeping track of the signal polarity for each test point with a dc voltage.

Such problems are less likely to surface when working with a DMM, since most digital multimeters (above the least-expensive models) automatically set the range, and a reversed polarity voltage simply results in a negative value on the display readout. Some DMMs also automatically determine whether the monitored signal is ac or dc and automatically switch themselves to the correct mode setting. With an analog multimeter (or other voltmeter), however, such things can be a major pain in the neck, and can all too easily result in a damaged meter.

A "quick-and-dirty" solution is to place a standard dc voltmeter (this can be your multimeter set up to measure dc voltages) in a Wheatstone-bridge configuration with four standard silicon diodes, as illustrated in Fig. 2-1. Almost any silicon diodes should

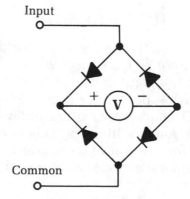

**Fig. 2-1**  *This circuit makes it easy to test voltages of any polarity, either dc or ac.*

work well in this application, although we recommend using the same type number diodes for all four legs of the bridge.

The bridge configuration of the diodes rectifies an ac voltage into a more or less equivalent dc voltage that can be read by the dc voltmeter. Moreover, if you reverse the polarity of the test leads when you are measuring a dc voltage, the diodes automatically compensate and correct the polarity as far as the voltmeter. No matter if the test signal is ac, positive dc, or negative dc, the voltmeter always "thinks" it sees a positive dc voltage.

This method is handy, but please don't expect miracles. This is a very crude trick, with a number of significant limitations. Do not use it in any application where precise voltage values are crucial.

Measurements of ac voltages are particularly likely to be inaccurate. You'll get the best results if the ac voltage is in the form of a sine wave. For ball-park voltage readings, however, this simple all-polarity voltmeter works well.

Even with dc voltages, there is some inaccuracy. The voltage drop across the diodes will be subtracted from the signal voltage before it reaches the voltmeter for measurement. At least, this effect will be fairly constant, and, if necessary, you can easily add a constant "fudge factor" to all voltage readings to compensate for the diode voltage drops.

Another limitation of this circuit is that it doesn't give you any indication of which type of voltage you are monitoring. You can't tell from the voltmeter if you have an ac voltage, a positive dc voltage, or a negative dc voltage. Still, this circuit is so simple and inexpensive, and can come in so handy for many practical testing situations, that this project is certainly worthwhile. Sure it's a "quick-and-dirty" trick, but sometimes that's all you need to do the job.

## Polarity indicators

Occasionally, when you are working on an electronic circuit, you might not be too sure of what type of signal you might find at a given test point. Is it ac or dc? If it is a dc voltage, then what is the polarity — positive or negative?

Of course, you can find out with a little careful trial and error, using your multimeter. This is a rather inelegant solution and calls for a lot of extra test-lead swapping and resetting of the multimeter's mode selector switch.

The simple circuit shown in Fig. 2-2 unambiguously tells you what type of voltage you are dealing with. A suitable parts list for this project is given in Table 2-1.

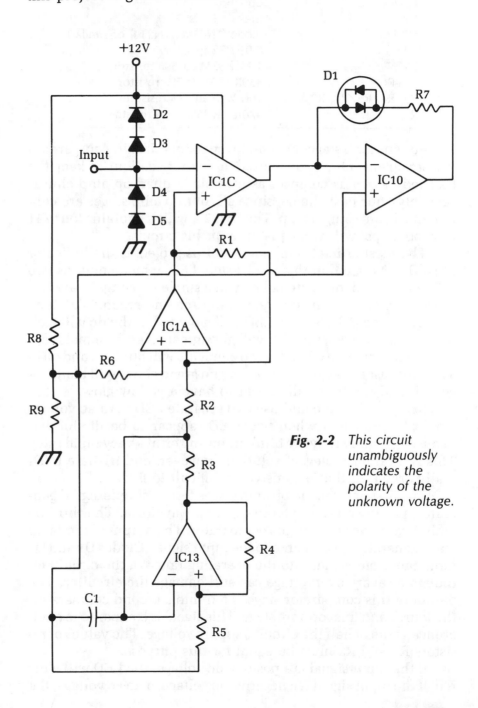

**Fig. 2-2** This circuit unambiguously indicates the polarity of the unknown voltage.

**Table 2-1  Parts list for the
polarity indicator circuit of Fig. 2-2.**

| | |
|---|---|
| IC1A–IC1D | op amp |
| D1 | tri-state LED |
| D2–D5 | diode (1N4148, 1N914, or similar) |
| C1 | 0.01 µF capacitor |
| R1 | 1 MΩ ½ Watt 5% resistor |
| R2, R6 | 330K ½ Watt 5% resistor |
| R3, R4, R5, R8, R9 | 10K ½ Watt 5% resistor |
| R7 | 470 Ω ½ Watt 5% resistor |

Four op amp stages are employed in this circuit. Any garden-variety op amp chips can be used in this application. To keep the finished device as compact as possible, a quad op amp chip is probably your best choice, since all four op amp stages are contained within a single chip. The LM324, which contains four 741 type op amps, will work just fine in this circuit.

The most crucial component in this project is the indicator LED (D3). Notice that this is a *tri-state LED*, which contains two LED elements of opposite polarity in a single housing. These two LEDs have contrasting colors, usually red and green. With a dc voltage of one polarity, the LED will glow red. If the dc voltage's polarity is reversed, the LED will glow green. If an ac signal is applied to the tri-state LED, the two elements will blink on and off at a rate too fast to be visible. The red and green glows will blend together, and the unit will appear to have a yellow glow.

In a pinch, you could use two separate LEDs. An ac voltage would be indicated when both LEDs appear to be lit simultaneously (actually they are blinking on and off at a very rapid rate). This is a rather inelegant solution, however, and tri-state LEDs really aren't all that expensive or difficult to find.

The first and second op amp stages form a simple signal generator, putting out a sawtooth or ramp waveform. The third op amp stage serves as a simple comparator. The output of the ramp wave generator is compared to the input signal. Diodes D1 and D2 force the input signal into the operating range of the circuit, although a severe overvoltage can still damage the circuitry. The output of this comparator stage is fed into a second comparator, the fourth and final op amp stage. This signal is compared to a reference value of half the circuit's supply voltage. The values of resistors R1 and R2 must be equal for this purpose.

If the input signal is a positive dc voltage, the LED will glow red. If the input signal is a negative dc voltage or an ac voltage, the

LED will glow green. Of course, if the LED doesn't light up at all, there is no signal (or a ground potential signal) at the test point.

# Full-wave power supplies

Figure 2-3 shows a full-wave rectifier power supply with a step-up power transformer. This type of circuit is used in many TV, PA, and radio circuits.

To read the supply voltage for the TV or amplifier (or the *load*), put the positive lead of the voltmeter on the filter output, which is one terminal of the choke (coil). Always set the meter to a scale that will show more than the voltage you expect to read. In typical TV circuits, this is between 300 and 500 V. Start with the meter on a 0 V to 500 V dc scale.

Notice one odd thing: while putting in 300 Vac at the transformer secondary, you get 350 Vdc at the filter input, after rectification. This apparent voltage increase is caused by a difference in the way the voltmeter reads ac and dc. Ac voltage is read as an *effective* or *rms*, which is only 0.707 of the actual peak or maximum

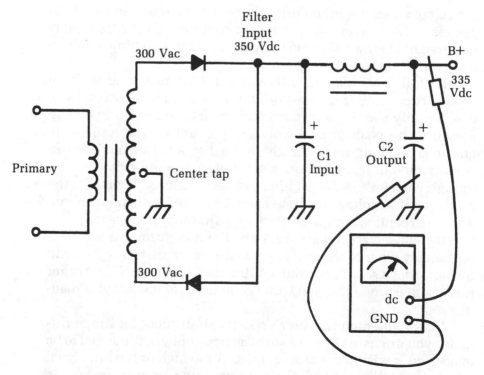

**Fig. 2-3**   *A full-wave rectifier power supply with a step-up power transformer is used in many TV, PA, and radio circuits.*

voltage in the waveform. (Conversely, the peak value is 1.414 times the rms value.) By rectifying the ac voltage, you get a series of dc pulses that reach the peak value of the input voltage. Here, this would be 300 × 1.414 or about 420 V.

When you feed this pulsating dc voltage to the large electrolytic input filter capacitor C1, it charges the capacitor to the peak voltage, less the "rectifier drop." The exact amount of voltage dropped across the rectifiers will depend on the characteristics of the specific devices used. For convenience, assume the rectifier drop is about 50 V. When drawing current out of the power supply to feed the load, this drops the voltage a little more, so you end up with about 350 Vdc at the filter input. Drawing the load current through the filter choke gives a small voltage drop, so you read about 335 Vdc at the filter output.

In some circuits, you'll find resistors used in place of the choke. The higher dc resistance makes the drop within the filter higher, so you get a lower voltage at the filter output.

# Measuring ac voltages

Ac voltage tests are inherently trickier than comparable dc voltage tests. The waveshape (harmonic content) and the frequency of the signal being measured both can affect the reading obtained on a multimeter.

If at all possible, ac voltages should be measured with an oscilloscope, rather than a multimeter. On an oscilloscope you can actually see the waveform and the frequency, and you can measure the peak to peak voltage by counting the number of granules taken up by the height of the displayed signal. You can get reasonable accuracy on your measurements this way, although you can't really get high-precision results. Typically, the accuracy of ac voltage readings made on an oscilloscope are about 3 to 5 percent, which is pretty good, although not great.

If you need greater accuracy and the waveform in question is reasonably close to a sine wave, you can measure the signal with the ac voltage section of your multimeter. However, if any other waveform is involved, you won't be able to get meaningful readings.

Check the manufacturer's specification sheet for the multimeter you are using to make sure the frequency of the signal to be measured is within the usable range. A too high or too low signal frequency will throw off the reading, often by a considerable amount. Some simple, inexpensive multimeters can't deal with frequencies much higher than about 40 to 60 Hz. But modern high-grade DMMs often can handle frequencies up to about 10 kHz (10,000 Hz) on the lowest voltage range. Of course, if the ac signal is riding on a dc voltage (not centered around ground potential), you will not be able to get an accurate reading from a multimeter.

If you need to measure a moderately low-frequency square wave, you can use the dc voltage section of your multimeter. The reading will be approximately one-half the actual (peak) voltage value. For example, a reading of 2.5 V would indicate a measured signal voltage of about 5 V.

This trick will only work with a fairly narrow range of frequencies. You will have to experiment to determine the useful limits of this test on your particular multimeter. If the signal frequency is too low, an analog meter's pointer will bob back and forth, making it very difficult, if not impossible, to read. On a DMM, you would just get a blur of continuously changing numbers, not any useful measurement. On the other hand, if the signal frequency is too high, you will get a steady reading, but it will become increasingly inaccurate as the frequency increases.

## Power supply ripple

Many modern electronics circuits require a fairly tightly regulated power supply. That is, the supply voltage should not vary much from its nominal value. Except for battery-operated portable equipment, most electronic devices are operated from the ac power lines through an ac-to-dc converter power supply circuit. Ordinary house current has a frequency of 60 Hz. All too often, some of this 60 Hz ac signal will manage to get through the power supply's filters and ride on the dc supply voltage, as illustrated in Fig. 2-4. This effect is known as *power supply ripple*.

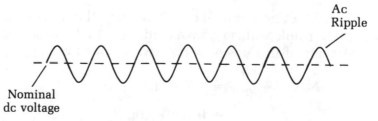

Ac
Ripple

Nominal
dc voltage

**Fig. 2-4**    *Some circuits are sensitive to power-supply ripple, which is an ac signal riding on the dc supply voltage.*

Of course, it is quite easy to spot such ripple with an oscilloscope, but when you are working on a power supply circuit, it probably is easier to use a multimeter. First, measure the dc voltage at the output of the power supply circuit, then switch your multimeter into the appropriate mode to measure ac voltages. *Note:* Please remove the test leads from the circuit under test or turn off the power to the system while switching modes. In some cases, your multimeter could be damaged if power is applied continuously to its test leads while you are switching modes.

Measure the ac voltage at the same test point where you took the dc voltage reading. This ac voltage will be roughly equivalent to the ripple signal riding on the dc voltage. It usually will be a very small value, unless there is something severely wrong with the power supply circuit. It certainly should be considerably lower than the dc voltage.

Divide the ac ripple voltage by the dc voltage and multiply by 100. In algebraic terms this is:

$$Eac/Edc \times 100 = \% \text{ ripple}$$

where $Eac$ is the ac voltage, and $Edc$ is the dc voltage.

As an example, let's assume the power supply is putting out a dc voltage of 12 V. The ac ripple voltage is 0.15 V, so the power supply's ripple rating in this case works out to:

$$\% \text{ ripple} = \frac{0.55}{12} \times 100$$
$$= 0.045 \times 100$$
$$= 4.5\%$$

This is quite a good reading and should be perfectly sufficient to power most electronic circuits adequately. However, some highly crucial circuits (such as some digital systems) might require even better regulation. You will need to add more filtering to get good results from the circuit using this power supply in this case.

Let's consider another example. This time the dc voltage is 9 V, and the ac ripple voltage gives a reading of 1.2 V. This means the ripple factor of this power supply works out to:

$$\% \text{ ripple} = \frac{0.2}{9} \times 100$$
$$= 0.133 \times 100$$
$$= 13.3\%$$

This much ripple is likely to cause some operational problems for all but the most noncrucial circuits. Assuming the power supply circuit wasn't just from a lousy design, there is probably a bad filter capacitor in the circuit. This is a fairly common type of problem, especially with older electronic equipment. It also can happen occasionally with equipment that has been left unused for an extended period. Large electrolytic capacitors are commonly used for power supply filtering. They can sometimes dry out and go bad if not occasionally refreshed by a suitable voltage being fed across them.

If the ripple factor is really outrageously large in proportion (say, about 25% or greater) to the dc voltage, suspect that one or more of the rectifier diodes in the power supply circuit has opened up or you have a bad transistor or voltage regulator IC. Such problems almost certainly will create severe misoperation of the circuitry being powered by the defective supply. In many cases, the result could be complete equipment failure. Fortunately, this type of problem is not too common.

## Shorted power transformers

A common source of problems in electrical circuits is the short circuit. The trick is in determining just where it is. It's usually somewhere in the load circuit, but not infrequently, the problem might be in the power transformer. You could waste hours going through every component of the load circuit and tearing your hair out because everything checks out fine, when the problem was really in the power supply itself.

A handy "quick-and-dirty" test is to disconnect everything from the secondaries of the power transformer. It should be driving no load at all. Now, plug in the device and wait a few minutes. Ten to twenty minutes would be appropriate. Feel the case. Is it abnormally warm? If it is, then the power transformer is probably shorted. If the case is too hot to touch comfortably, then there is definitely a problem with the power transformer.

You've narrowed down the problem to the power transformer, since nothing else is getting power, so no other component can be the source of the heat. The transformer cannot be overloaded, since it is running with no load at all, so the current drawn from its secondaries should be zero. The only possible cause is some shorted windings within the transformer itself. The transformer must be replaced or rewound.

You can use a dynamometer-type wattmeter for a more exact test. A good power transformer with no loads should give a very small reading, of 5 W or less. This small wattage is caused by the normal iron-loss effects within the transformer. A higher wattage reading with no load is a clear indication of trouble.

Not all wattmeters will work properly for this test. You must use a dynamometer type, with both a voltage coil and a current coil. Such a wattmeter will have four terminals, instead of just two.

What if you don't have a suitable wattmeter handy? You can make a simple, crude, but effective wattmeter with an ac voltmeter (the ac voltage section of your multimeter) and a 1 Ω resistor in series with the ac circuit to be monitored.

# Power supplies with voltage dividers

In actual circuits, you need to divide the V+ voltage to get the proper values on the different circuits. Therefore, you must use a circuit called a *voltage divider*. Such a circuit is shown in Fig. 2-5. With the V+ voltage from the full-wave power supply just discussed, all we need to do is make the + and − connections to the divider.

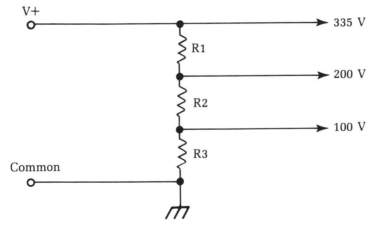

**Fig. 2-5**  *This power-supply circuit includes a voltage divider to supply different voltage values to various portions of the load circuit.*

When hooking the series of resistors across the supply, you get a small, constant current flow in the resistors. This bleeder current helps to stabilize the voltage by furnishing a constant load on the supply. You can tap off any voltage you need by tapping down on the voltage divider.

You'll find a lot of variation in resistance values used in voltage dividers, but the total will probably be somewhere around 40 to 50,000 Ω. The total resistance is chosen so the bleeder current will not be too high, which could overload the power supply. The values of the individual resistors are chosen to divide the total voltage as desired.

To check the resistors, put the negative lead of the voltmeter to ground, and read the voltages at the taps. For instance, you might get something like this: The 335 V tap reads 345 V; the 200 V tap reads 210 V; the 100 V tap reads 0. What's the trouble? R2 is open. Note that the taps that still read voltage have higher readings than normal. This means that part of the normal load has been lost. The bleeder current has also ceased, since the circuit from which V+ to ground is open, which raises the voltages still more. Double-check by turning the set off and checking the resistance of R2 with an ohmmeter.

Now let's see what happens if we find *almost* the same symptoms as in the preceding unit. Suppose the 335 V tap reads 320 V, the 200 V tap reads 180 V, and the 100 V tap reads 0 again. Note that these voltages are now lower than normal. You'll notice another typical symptom if you touch the resistors: R1 and R2 are now very hot, and R3 is cold.

If you turn the set off and take resistance readings, all of the resistors themselves are correct; however, when you read the resistance from the 100 V; tap to ground, there's a 0 reading. There is the trouble! Check the bypass and filter capacitors in the circuit; one of them has shorted out.

The key clues, other than the overheating of the other resistors, are the low voltages. When the capacitor shorted, it took R3 out of the circuit completely. As a result, there is that much less resistance from V+ to ground, so the bleeder current will go up and the total voltage will go down. In many cases, this kind of overload will be enough to burn out R1 or R2. So if you find an open resistor at any point in a voltage divider circuit, check on its *load* side (the end away from the source of voltage) to make sure that there is no short circuit there that could burn out a newly installed resistor.

# Power supplies with dropping resistors

Figure 2-6 shows the equivalent of the previous circuits, but with the resistors arranged a little differently: individual resistors are connected into each voltage-supply circuit. The result is the same — each circuit gets its proper voltage. Without the multiple-

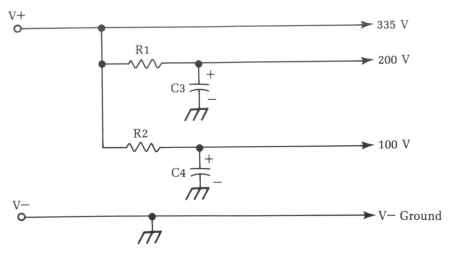

**Fig. 2-6**  *This power-supply circuit features dropping resistors.*

tapped voltage divider resistors, there is no bleeder current in such a circuit.

The dropping resistors are chosen so that their values will give the correct voltage to each circuit. This feature gives a very good clue when there's trouble. If the voltage on any tap is low or high, then check that circuit first to see why. The rest of the voltages won't be affected as much, but will be changed to some extent.

First check the supply voltage (this rule holds for all power-supply troubleshooting). Without the right supply voltage, the other voltage measurements are meaningless. Use some of the same obvious clues as before. For example, if some of the dropping resistors are very hot and the supply voltage is low, you know immediately that there's a short circuit somewhere.

If the supply voltage is low, but there are no signs of overheating, then look for something that is weak — a leaky or shorted diode, an open input filter capacitor, a weak selenium rectifier, or anything that would reduce the ability of the power supply to deliver the right amount of current.

These tests apply to all circuits. In the circuit shown, a short in one load circuit will kill the voltage at that tap, but it won't make the others drop as much as before. For example, if C3 is shorted, the 200 V line will read 0, the 335 and 100 V lines will go down by about 10 percent. The key clue will be R1; it will be very hot. If the short has existed long enough to cause R1 to burn out, then the 200 V line will read 0, but the other voltages will be above normal because of the loss of the normal load current in the 200 V line.

Many sets use this circuit. The average dropping resistor will be a 2 W carbon type. If it's been overheated, you'll notice a decided change in its appearance. The case will be darkened and the color-coding paint will have changed color because of the heat. The red bands, if there are any, are particularly valuable as indicators. They are the first to show a change of color from overheating and usually turn a dark brown.

Check any resistor that shows signs of overheating. In fact, it's a good idea to replace it on general principles because the overloading might have changed the resistance value.

## Power-supply circuits with branches

Figure 2-7 shows a power-distribution circuit that can be found in all types of electronic equipment using either transistors or tubes. The main differences between the tube and transistor circuits are in the resistor values and the voltages. Voltages typically range from 30 V to 40 V for transistors and up to 300 V or 400 V for tubes.

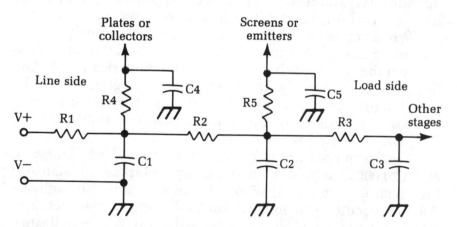

**Fig. 2-7**   *A power distribution circuit like this one is used in many types of electronic equipment.*

In Fig. 2-7, the supply point (V+) is again at the left-hand side, V− is ground. The current goes through R1, then branches through R2 and R4; the current through R3 goes on to other circuits. Each junction point is bypassed by a capacitor which is absolutely necessary to keep the RF impedance of the power supply very low. Since this distribution circuit is common to all stages, it must prevent any interaction between the various circuits. If not, you get feedback and the equipment will oscillate.

The capacitors are the most common cause of voltage troubles. If any of them shorts, it will kill all voltages past the point where it is connected in the circuit. However, there's a quick and easy way to find a shorted capacitor.

In Fig. 2-7, V+ is the supply side or line side of the network. The right-hand side of the circuit is the load side — the stages that need the voltage. If you know where the current is coming from and what it goes through on the way, you can tell what's wrong when it doesn't get to where it should!

For instance, if C2 shorts, you'll read 0 V at the junction of R2, R3, and R5. Clue No. 1: Voltage is present at the junction of R1, R2, and R4, but it isn't as high as it should be. Clue No. 2 is easy to spot: R2 is probably getting very hot. R3 and R5 are cold. So look on the load side of R2 for the short. The current must go through the last hot resistor in the circuit before it gets to the short. Turn the power off and take an ohmmeter reading. Look for a zero resistance to ground; such a reading is always incorrect in a voltage supply circuit, so there's the trouble. To confirm this diagnosis, disconnect C2 and check it. The short disappears from the resistor network, and C2 shows a high leakage or short.

When you're hunting for a short in the power supply, start at the power supply (V+) and go from there toward the load(s). Remember this procedure because it's always the easiest way to find the cause of the trouble. Use the schematic diagram. Trace the power-supply circuits to each stage through the dropping resistors past any bypass capacitors that could short out until you get to the point where there's no voltage.

Transformer-powered circuits have normal resistances of about 20,000 Ω to ground. This is mainly the leakage resistance of the electrolytic filter capacitors. In silicon-rectifier or voltage-doubler circuits, you might have to disconnect the rectifiers. They offer a very low back resistance that can falsely indicate a short circuit.

Another quick check is to simply disconnect some of the loads and see what happens. If the supply voltage suddenly jumps back to normal, the last load disconnected could be the trouble spot. Also, you can make mental additions of the various dropping resistors in the circuit by checking their values on the schematic; take resistance readings at each end of the circuit and compare. In Fig. 2-7, for instance, if all resistors were 1,000 Ω, and C1 was shorted, you'd read 1,000 $\Omega$ to ground from V+ and 2,000 $\Omega$ to ground from the load end of R3 (R2 + R3).

# Transformerless power supplies

Many types of equipment now use transformerless power supplies for economy. Instead of stepping up the ac voltage with a transformer, apply the line voltage directly to a rectifier at the standard value of 117 V rms. Actual line voltage might vary, of course, but most equipment is designed to work, as stated on the rating plate at "105 to 120 Vac."

Figure 2-8 shows the simplest possible circuits: a half-wave rectifier and filter circuit (the filter is the same as before). In this circuit, again, the dc voltage is higher than the ac input. With a 117 Vac input, there's about 145 V at rectifier and about 135 V at the filter output. The input capacitor (C1) will be a large electrolytic, usually from 60 to 100 microfarad ($\mu$F). A resistor of about 1,500 $\Omega$ is used as a choke.

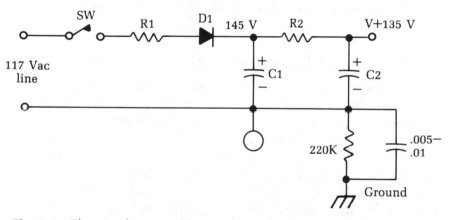

**Fig. 2-8** *The simplest possible transformerless power-supply circuit is a half-wave rectifier and filter.*

To measure dc voltages in this circuit, take the readings between V+ and A−. Measurements between V+ and ground won't be accurate because of the large series resistor. One fast way to find the negative terminal is to put the voltmeter negative prod on the negative terminal of the electrolytic capacitor. In can-type electrolytics (with one exception) the can is always negative. In cardboard-tube types, the black wire will be the negative if standard color coding is used.

# Half-wave voltage doublers

The 135 V from a simple transformerless power supply isn't high enough for TV circuits, so a circuit is needed that will give a higher voltage. Such a circuit is the voltage doubler, shown in Fig. 2-9. There are several types of voltage-doubler circuits, but this one is the most popular, probably because it's the simplest. It is called a *half-wave voltage doubler* because it uses each half of the incoming line voltage to charge a separate capacitor. The two capacitors are then discharged in such a way that the voltages add up. This is the basis of all voltage-doubling circuits.

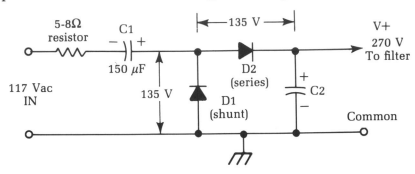

**Fig. 2-9**  *A voltage doubler is used to obtain a higher supply voltage.*

Although we are told that an electrolytic capacitor should never be connected to the ac line, here we seem to have one hooked right to it: C1 is in series with the ac input. However, following the return circuit, you can see that the capacitor is never subjected to a true alternating voltage. On the first half-cycle of voltage, the polarity is such that D1 is conducting. Current flows into C1, then out through D1, which is the "shunt" rectifier, and back to the other side of the ac line (note that this is a unidirectional current). When we say that current flows through C1, we mean only this one-direction *charging* current. Such a current flows in all capacitors while they are charging to an applied voltage.

Now, C1 is charged to about 135 V, and the next half-cycle of line voltage has such a polarity that D1 is not conducting, but D2 is. So, the current from the line flows on through D2 and charges C2; at the same time, C1 discharges through the same circuit, and its voltage is added to the charge on C2 because the polarity is the same (positive with respect to B—), C2 charges up to double the line voltage and you get about 270 V instead of the 135 V that was in the ordinary half-wave rectifier circuit.

The voltages here are read to an isolated common point, as shown. The most important thing to look for in such voltage readings is balance. If both rectifiers and both capacitors are good, you will be able to read the dc voltages as shown: 135 V across D1, from common to the center tap of the two rectifiers, and an equal 135 V directly across D2, from center tap (place the negative meter lead here) to positive of V+. If you read 135 V across one rectifier and only about 80 V across the other, the chances are that the "low" rectifier is faulty.

If the input capacitor loses capacitance or opens, the dc voltage will drop to less than half normal or disappear entirely. If C2 opens or loses capacitance, the dc voltage will drop quite a bit and there will be a great increase in the hum level.

The small resistor seen in the ac input is called a *surge resistor* and is a fusible type; although it is a wirewound resistor of about 4 to 5 watts in rating, it is designed to act as a fuse. If something shorts out in the power supply, it will open and prevent further damage. (A good check for this resistor in actual circuits is to carefully feel it after the set has been on for a minute or two. If it's warm, it's okay. If it's "stone cold," it is very apt to have been blown.)

This is the instance mentioned earlier where the can or negative terminal of an electrolytic capacitor is not common. Notice from the circuit that *both* sides of C1 are "hot" with respect to the common, and practically everything else!

Because of the pulse nature of the current and voltage across C1, it is very difficult to check this capacitor with a voltmeter. The symptom of an open C1 is a large drop in V+ voltage; the quickest test for this condition is to shunt a known good electrolytic capacitor of about the same size across it. Hook your dc voltmeter to the output with test clips, note the reading, and then shunt the test capacitor across C1; if the dc voltage jumps up to normal, replace C1. You don't have to use the same size capacitor for testing. If you replace a capacitor, always use exactly the same size; but almost any size capacitor will do for a quick test. (If C1 is 150 $\mu$F, as many are, you can shunt, say, an 80 $\mu$F unit across it; a very definite rise in the dc voltage indicates that the original capacitor is bad.)

# Full-wave voltage doublers

The full-wave doubler, illustrated in Fig. 2-10, is the original voltage-doubler circuit. It was used with special dual-diode vacuum tubes in old radios and is now found in quite a few circuits with

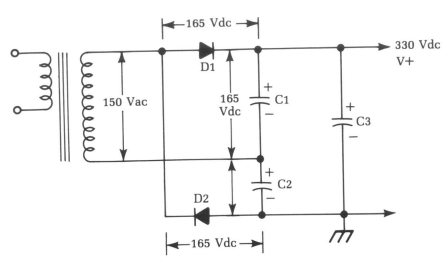

**Fig. 2-10**   A full-wave voltage doubler circuit.

selenium or silicon rectifiers, as shown here. To make the circuit easier to read, it's shown fed by an isolated transformer secondary winding—a circuit that you'll find in actual equipment in many cases. In a few cases, this circuit is used without the isolation transformer—not too many, though, because the floating center tap makes it hard to handle with respect to the rest of the circuits.

It works like this: On the first half-cycle of voltage from the ac supply, D1 conducts and charges C1, just as in the half-wave rectifier circuit. Because D2 is reverse-biased during this time, it does nothing. On the next half-cycle, the polarity reverses; D2 conducts and charges C2, and D1 is cut off. Now two big electrolytic capacitors, each one charged to 165 Vdc, are connected in series. So, you can take the "added" voltages off and feed them to the filter as the sum, or 330 Vdc. The action of this circuit depends on the alternate-charging and series-discharging of these two capacitors. This is the basis of all voltage-doubling circuits, but it is not so apparent in the others. Here, you can see the two separate capacitors.

Tests are the same as in the half-wave circuit. Read the dc voltage across each rectifier, watching your voltmeter polarity; the reading should be 165 Vdc on each one. The same goes for capacitors C1 and C2: the voltages must be equal because the capacitors are to be the same size. The dc output voltage is the sum of the voltages across the two capacitors. C3 is the filter capacitor.

If one of the doubler capacitors should lose capacitance or open, the circuit will not balance and the output voltage will drop quite a bit. The same thing happens if one of the rectifiers goes bad. As mentioned, the doubler capacitors must match; you'll find sizes from 40 $\mu$F to perhaps 200 $\mu$F used here. The size of the doubler capacitors affects the developed dc voltage. As a rule, the bigger the doubler capacitors, the higher the dc voltage, because larger capacitors will hold more charge.

The filter capacitor, C3, has nothing to do with the doubling action; similarly, the doubler capacitors have very little to do with the filtering. So if you have a bad hum with almost normal output dc voltage, the filter capacitor is weak. If you have some hum and the voltage is low (more than 25 percent), one of the doubler capacitors has probably gone weak or open.

In the isolated circuit shown in Fig. 2-10 the common point can be grounded to the chassis. In line-connected circuits without any isolation, both the capacitor center tap and common must be isolated from the chassis, as in the half-wave circuit. This can be confusing when you are taking dc voltage readings, so watch out. Always connect the negative lead to the voltmeter to the negative terminal of the filter capacitor, not to one of the doublers, unless you are sure that you are on the negative terminal of C2.

## Full-wave bridge rectifiers

Figure 2-11 shows a common circuit you'll find in color TV, two-way FM transmitter-receiver power supplies, and many other applications, including low-voltage supplies for ac power of transistorized equipment. This circuit is a full-wave bridge rectifier. It has many advantages. Notice that the isolation transformer

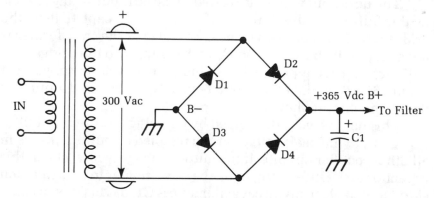

**Fig. 2-11**   Full-wave bridge rectifier circuits are used in many types of electronic equipment.

needs no center tap. So, the full secondary voltage is available at the rectifier output, rather than half of it, as in the center-tapped full-wave rectifiers seen previously. This arrangement saves both size and cost in the power transformer. The circuit can be made fully isolated, and B— is usually the chassis.

On one half-cycle, one end of the secondary winding is positive. The other end is negative. Current flows through D1 and D4 to the load and ground, respectively (in the same direction with respect to the load), while the other two rectifiers reverse polarity and do not conduct. On the next half-cycle, D1 and D4 reverse-bias and cut off, and current flows through D2 and D3. So, it uses both halves of the cycle, as in the full-wave circuit shown earlier.

One advantage of the bridge circuit over other circuits is that there are always two rectifiers in series across the total ac voltage of the secondary. Thus, for a given output voltage, the individual rectifiers have a lower peak voltage applied to them than in other circuits. As a result, they have longer life expectancy, can use lower ratings, etc. The ripple frequency is 120 Hz, as in other full-wave circuits.

When making replacements, look for the identification on the rectifier. Notice that there are two positives hooked together and two negatives. They are always B+ and B—, respectively. Each end of the transformer winding always gets one positive and one negative (it doesn't matter which of these is which). In some circuits, you might find bias resistors connected between B— and chassis ground. They work in exactly the same way as in the previous full-wave circuits.

The dc output voltage is related to the ac input voltage as it is in the full-wave tube-type rectifier circuit, except that in the bridge circuit the entire secondary voltage of the transformer is the ac input. If you have a 300 V rms ac supply to the bridge, you will read the same proportion of increase as before; so, you get about 365 Vdc at the filter input, read between B+ and ground or B— in the circuit shown.

This dc voltage is developed by charging C1, the input filter capacitor, to the peak voltage of the rectified ac output, just as in all other rectifier circuits. If the output voltage is low, check this capacitor first. Once again, hook the dc voltmeter to B+, and then shunt a good electrolytic capacitor across C1. If C1 is low in capacitance, you'll see the voltage jump back up toward normal. If C1 is completely open, your dc voltage will probably read about 40 to

50 percent of normal, or even less, depending on the load current being drawn.

In solid-state circuits, a special bridge rectifier is often used. The four diodes making up the bridge are packaged in a single unit. Such devices function exactly the same as separate diodes.

# Electrolytic capacitors in series for higher working voltage

Figure 2-12 shows the basic schematic of a circuit that you will find now and then in high-powered amplifiers and in a few transmitters. Notice that the filter circuit has a choke input for better voltage regulation. This choke input cuts down on the peak voltage but helps to hold the output voltage much steadier.

In this circuit, we're getting into V+ voltages that are above the normal working voltage of electrolytic capacitors. The average TV filter capacitor is rated at about 450 working volts, with a 550 V peak surge rating. This means that the capacitor should withstand 450 V—at turn on, for example—without blowing out. This peak rating shouldn't be exceeded or even reached for more than about 15 to 20 seconds. The working-voltage rating, for the best results and longest life, shouldn't even be met. A 450 V capacitor, for example, shouldn't be used in circuits with steady or normal voltages of more than about 400 V, and this is cutting it close. A 450 V capacitor working at about 350 V is about normal in the average well-built TV set or amplifier.

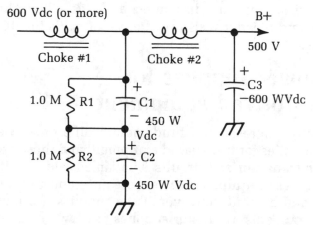

**Fig. 2-12**  *This circuit is used in some high-power amplifiers.*

There are times, however, when you must have an electrolytic that will hold a higher voltage, so put a couple of capacitors in series. This method works just as with paper capacitors; you wind up with half the capacitance and twice the voltage rating. This scheme is used in Fig. 2-12. Here, C1 and C2 are 450 V capacitors (900 V total) and big enough so that the resultant capacitance is enough to give adequate filtering. In a typical commercial circuit, a pair of 100-$\mu$F capacitors is used, giving 50 $\mu$F at 900 V.

One thing to remember whenever such units are replaced is the two capacitors must be of exactly the same value for best results. If you use 100 $\mu$F and a 60 $\mu$F in series, the smaller capacitor will assume a greater percentage of the total voltage and probably blow! If you have to replace one capacitor of a pair like this, always use an exact duplicate. Actually, if one shorts out, replace both because the remaining capacitor has been severely overloaded and could fail in service soon.

The 1 M resistors hooked across the capacitors are necessary to equalize the voltages. They take no perceptible current from the circuit because they are so big. They only draw 1 mA of current for each 1,000 V across 1M. So if we had 600 V here and 2 M, our total current would be 0.3 mA (not enough to make the power transformer even run hot.)

Quick check in this circuit: Read the dc resistance from V+ to ground at the rectifier cathodes or filaments. If you read 1 M, one of the filters is shorted. (You're reading a short through one and a 1 M resistance across the good one.) The normal reading, of course, would be 2 M or less, depending on your ohmmeter polarity. You'll read a certain amount of leakage resistance through the electrolytics. You must disconnect all loads—including voltage dividers and bleeder resistors, if any—for this test to have any meaning.

# Ac power supplies for transistorized equipment

Essentially, there are no fundamental differences between ac power supplies for tube-based equipment and those designed for use with transistorized circuits. The same basic circuits are used in both types of equipment. There are differences in the values of voltage and current, however. Tube supplies use high voltages and low currents; transistor supplies use low voltages and high currents. If a TV supply has 350 V output at 250 mA, it draws

87.5 W. A transistor amplifier power supply with a —25 V output at 3.5 amps also draws 87.5 W. There is no difference in the power supplied; only the values of voltage and current are different. As long as their product stays the same, the output power is the same.

In one respect, you need different test equipment for transistor circuits. In tube circuits you measure high voltages; in transistor circuits you often need to measure very small voltages — frequently 0.1 V or so. With tubes, voltage *tolerances* can be very high. If the rated plate voltage of a tube is 100 V, in many cases it can vary from 90 V to 110 V without making a lot of difference in the output of the tube. (In some circuits, tolerances are as high as 20 or even 25 percent.) Hence, your dc voltmeter can have a fairly large error before it makes any difference in a tube circuit.

Transistor voltages are more crucial for two reasons. First, the levels are only in the range of tenths of a volt; and second, tolerances are much tighter than with tubes. In the bias voltage reading of a typical transistor amplifier, you might find such values as base, 0.4 V; emitter, 0.6 V — giving a 0.2 V difference. Trying to read that on a 5 V range (the lowest on many VOMs) is demanding a great deal from an instrument whose accuracy at full scale might be no better than 5 percent. In the lowest quarter of the scale, it could be considerably poorer. A good VOM for this kind of work should have a 2.5 V (or lower) low range with at least 20,000 Ω per-volt sensitivity and 2 percent full-scale accuracy.

Changes of a fraction of 1 V can come from a variation in the supply voltage. Voltage-regulated power supplies are used quite often in better transistor equipment. Like tube circuits, the cheaper transistor units simply use "straight" power supplies and depend on the regulation of the power line to keep them within limits. Even the best circumstances, however, the ac line voltage varies constantly as its load changes. The use of some kind of voltage regulation makes transistor equipment work much better. Batteries, of course, furnish a steady source of power. In high-quality equipment, such as laboratory-type transistorized voltmeters, even the battery supply goes through a voltage regulator!

Transistors are far more sensitive to hum than tubes. Transistors operate on very small voltage changes, which cause very large current changes for a given power output in watts. Since hum is always a voltage phenomenon, you really have to lean over backward to make sure you have taken out every last bit of hum and ripple in the power supplies. Some highly specialized circuits in the next few pages have done so. Voltage regulation serves a dual purpose here; since hum is voltage variation, you automatically take out the hum if you use a regulator that holds the voltage absolutely constant.

# Power supplies from a transformer winding on a phono motor

A simple power supply circuit found in a number of small phonographs with transistor amplifiers is shown in Fig. 2-13. The novelty is in the use of a separate winding on the phonograph motor, making the motor serve as a power transformer as well. Transistors need low voltages but high currents. So using a power transformer to obtain a suitable supply voltage is much better than the only other method: a dropping resistor in series with the ac line. With a dropping resistor, heavy current changes that occur in transistor circuits would cause the supply voltage to change as well, upsetting the circuits.

In the circuit in Fig. 2-13, a simple half-wave rectifier is used. In others you might find full-wave rectifiers, bridge rectifiers, and so on. They all work on the same principle as the power supplies discussed earlier.

In a variation of this circuit, used with small one-tube phonograph amplifiers, the filament of the tube is connected in series with the phono motor. The motor, of course, must be specially wound to work on the lower voltage. A typical example of such a

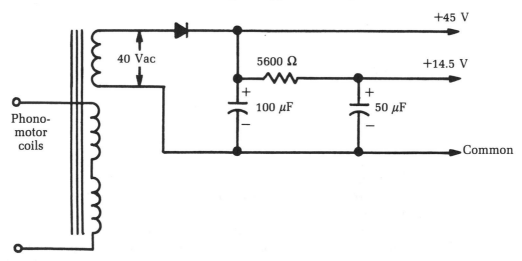

**Fig. 2-13**  *In some small phonographs, the motor is used as part of the power-supply circuit.*

circuit uses a 25 V tube in series with a 92 V motor; 92 + 25 = 117 Vac, so the combination can be connected directly across the line. You can always tell by its behavior when this circuit is used; if the tube blows out, the motor stops!

# Transistorized capacitance multipliers

Transistor circuits need very good power-supply filtering. They also need very good voltage regulation, which is another way of saying the same thing. Very large capacitors help stabilize supply voltages; their charge furnishes extra current when sudden peaks occur in the load current, holding the voltage steady. Even with the large capacitances used in transistor circuits, you can always use more. Figure 2-14 shows a voltage regulator circuit called a *capacitance multiplier.*

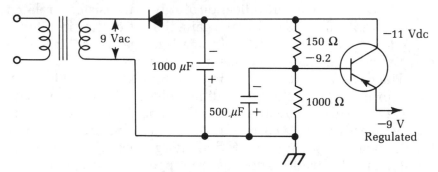

**Fig. 2-14** *This circuit is known as a capacitance multiplier.*

The load current from the power supply flows from the collector to the emitter of the transistor, exactly as in the voltage-regulator circuits. You control the transistor's resistance with its base bias. You can make the resistance higher for low currents and lower for high currents, thus holding the output voltage steady.

Notice the voltage divider connected between the —11 V supply and ground. The transistor base is hooked to the junction of the two resistors. Under normal load, the base will have a certain voltage (—9.2 V here) that will hold the emitter voltage constant at —9 V. It does so by controlling collector-emitter current or, looking at it another way, by controlling the resistance of the transistor.

If the load current goes up, the bias changes in response. The higher current loads the power supply; the supply voltage drops. The voltage across the voltage divider drops, and the base bias drops with it. This change makes more current flow through the transistor (lowers the resistance); so, there is now a lower voltage drop across the variable resistor and the output voltage is held where it was. (This action occurs in a split second.)

In effect, the circuit uses the 2 V drop across the transistor as a reserve. When the supply voltage tries to go down, the circuit adds part of this reserve to the output voltage. The reverse is also true; if the supply voltage tries to go up, the transistor takes on a larger surplus.

Why call this circuit a capacitance multiplier? Because of its effect. If you had a very large capacitor across the power-supply output, it would hold a quantity of electrical energy; from this reservoir it could release energy whenever needed to hold the output voltage constant. The capacitance multiplier does the same thing with a smaller capacitor.

The capacitance multiplication of a given transistor is related to its current gain: the higher the gain, the better the circuit will work. (Current gain is the ratio of a small change in base bias to the corresponding change in collector-emitter current.)

The base capacitor (500 $\mu$F in the circuit shown in Fig. 2-14) plays an important role. By holding a charge, it holds the base voltage as steady as possible. When the supply voltage drops, the base voltage would also drop if it weren't for this capacitor. The base capacitor discharges part of its energy through the 1,000 $\Omega$ resistor, and the resulting voltage drop opposes the supply voltage drop and holds the base at the normal voltage.

# Checking capacitance multiplier circuits

To check a capacitance-multiplier operation, check all the dc voltages and compare them with the values given on the schematic diagram. They should be very close. Be sure to check the ac input voltage to see that it is normal; most circuits are rated at 117 Vac input. Vary the loading. In an audio amplifier, for example, turn the volume full up for maximum load and then all the way off. This variation should have no effect on the output voltage of the capacitance-multiplier circuit.

If the output voltage is low, disconnect the load. If the voltage jumps back up to above normal, there could be a heavy overload of current being drawn by the amplifier — more than the regular circuit can handle. The load current can be measured by putting a milliammeter in series with it at the voltage-regulator output. The proper value will be given on the schematic.

Check the dc voltage between the collector and emitter of the regulator transistor. If this difference is very low, or zero, check the base voltage before you decide that the regulator transistor is shorted. If, in the circuit shown in Fig. 2-15, the base should be down to about −8 V instead of the normal −9.2 V, the transistor would be conducting as hard as it could in an effort to hold the voltage up. A wrong base voltage could result from changes in the values of the voltage-divider resistors, excessive leakage in the base electrolytic capacitor, etc.

As usual, an alternate trouble can show similar symptoms. Check the control transistor for shorts and leakage. If you're not certain, take it out and check it with an ohmmeter; all transistors should show the diode effect on each pair of elements. Also, check the schematic to see if there is a shunt resistor connected across the control transistor, as shown by the dotted lines in Fig. 2-15. Such a resistor is used in some circuits to reduce the current loading on the transistor junction. It is a fairly low-value resistor and of course would upset any in-circuits tests for back resistance between collector and emitter leads.

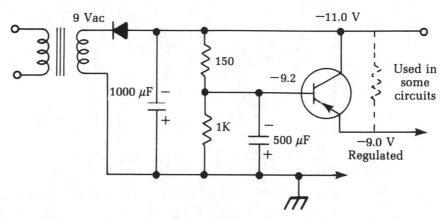

**Fig. 2-15**    *In some cases, a shunt resistor (shown in dotted lines) is used in a capacitance multiplier circuit.*

If the output voltage is low — say, by about half the normal voltage — check the input filter capacitor. If this capacitor is open, there will be no reservoir effect, and the output voltage will drop drastically. It is not a good idea to bridge capacitors for test in transistor circuits; the charging pulse can cause transients that might damage transistors. For safety, turn the set off, clip the test capacitor across the suspected unit, and then turn the set on again. With the instant warmup of all transistor circuits, you won't lose any time.

If you suspect the rectifier of being weak, use the same procedure. You can clip another rectifier across the original unit—watch the polarity—and turn the set back on. If this action brings the voltage back up, replace the old rectifier. Silicon rectifiers will not get weak as seleniums do; they usually short out completely when they do fail. You might find one that is completely open, but not often.

# Transistorized series voltage regulators

Although the circuit illustrated in Fig. 2-16 is called a transistorized series voltage regulator, it bears a strong resemblance to the capacitance multiplier just discussed. The only real difference is in the polarity of the voltages. An npn transistor is used here, since the output voltage is positive with respect to ground. The basic action is the same, though, with a few refinements. The collector-emitter "resistance" of the transistor is controlled by the base bias to keep the output voltage at the same level under varying load conditions.

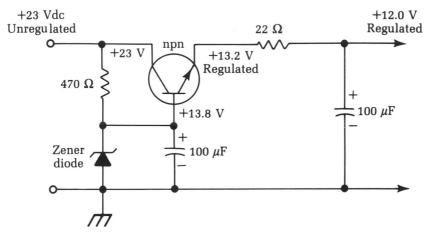

**Fig. 2-16** *A series voltage regulator circuit is quite similar to the capacitance multiplier.*

A zener diode is used to hold the base voltage constant. A zener diode acts as a normal diode when reverse-biased until a certain voltage is reached. Then it breaks down and carries a very heavy current. This breakdown is sometimes called a *controlled avalanche.*

When this conduction occurs, the voltage drop across the diode remains essentially the same over a wide range of current flow. (A protective resistor in series is used to hold the maximum current within safe limits and prevent complete burnout of the diode.) This variable-current-constant-voltage action is the same as that of gas-filled voltage-regulator tubes.

Zener diodes come in many different voltage ratings, which are determined by their construction. To establish a certain reference voltage level, simply choose a zener that has that voltage rating. In the circuit shown, as you can see by the base voltage, the diode regulates at 13.8 V. A 100 $\mu$F electrolytic filter capacitor helps the zener out a little.

The 22 $\Omega$ resistor and 100 $\mu$F electrolytic at the right of the figure are not a part of the regulator circuit; the resistor drops the 13.2 V regulator output to exactly 12 V for use by one of the circuits. The electrolytic capacitor is a filter and a bypass capacitor as well, to keep signals from mixing in the power supply and causing feedback or oscillation.

## Transistorized shunt voltage regulators

There is one more transistorized regulator that we should mention. You won't find it used often because its efficiency isn't as good as that of the series-type regulator. The circuit, called a *shunt regulator*, is shown in Fig. 2-17. It is simple but it tends to waste power.

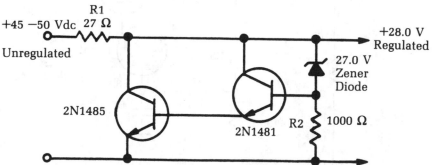

**Fig. 2-17**   *Some voltage regulator circuits use a shunt, rather than a series hook-up.*

The operating action of the shunt regulator is the opposite of that of the series type. The regulator transistor (or, in this case, the pair of transistors) is connected across the load in parallel, instead of being in series with it. If the supply voltage goes up, the regulator draws more current, instead of less. This extra current increases the voltage drop across R1 and brings the voltage back down. The power consumed by the shunt transistor(s) and R1, then, can be said to be wasted.

The operating principles of series and shunt regulators are similar. The zener diode clamps the base voltage of the transistor. The voltage drop across series resistor R2 is used to signal the regulator when more of less current is needed.

# Three transistor voltage regulators for transistor TV sets

Transistor TV sets need very close regulation. The three-transistor circuit shown in Fig. 2-18 is used by at least one leading manufacturer. This regulator uses the same principle as the single transistor circuit presented earlier, but this one is more elaborate. A pair of direct-coupled transistors is used as the regulator, controlled by an error amplifier. A 47 Ω resistor is shunted across the voltage-regulator transistor from the collector to emitter to carry part of the load.

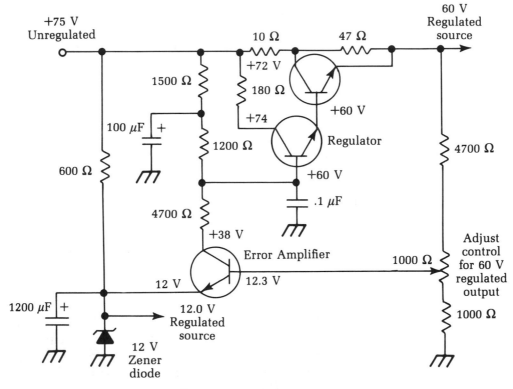

**Fig. 2-18**  *A somewhat more elaborate voltage regulator circuit.*

The error amplifier makes this circuit quite sensitive to small load-voltage variations. The base is connected to the slider of the voltage-adjust control, which is a part of the voltage divider across the output. In the emitter circuit, a 12 V zener diode clamps the voltage at this value and also provides a 12 V regulated source for some of the TV circuits. The collector is supplied from the 75 V line through the three resistors; it is also connected to the base of the lower transistor of the regulator. So, the collector voltage of the error amplifier sets the base voltage of the regulators.

If the load current goes up, the output voltage goes down — that is, more negative. The base of the error amplifier also goes more negative. This transistor is an npn unit, so the shift in base voltage changes the bias so that *less* current flows in the collector-emitter circuit. This circuit includes the three resistors from the 75 V unregulated source. With a lower current flowing through these resistors, the collector voltage goes more positive (it rises), thus making the base of the lower control transistor more positive. It's another npn, so the positive-going base voltage makes it draw more collector current. This reaction in the lower transistor is transferred and amplified by the upper one; its collector-emitter current increases (or its resistance decreases) and more current is fed to the load circuits, making the voltage rise again. If the opposite happens and the load decreases, making the output voltage rise, just repeat all of this but reverse the polarity of each reaction.

This regulator is adjustable. Using a Variac or adjustable transformer, set the ac input at exactly 120 V, and connect an accurate dc voltmeter across the 60 V regulated output. With the TV set tuned to a station, adjust the voltage control to make the dc meter read exactly 60 V, and that's all there is to it. To check operation of the regulator, vary the input ac voltage up and down 5 V to 8 V from the 120 V level. The dc output voltage should stay at 60 V.

If this test shows too much variation, check the dc voltages on all three transistors. See that the collector voltage of the error amplifier changes as you move the voltage-adjust control. If it doesn't, go back and check the base voltage of this transistor to see if it is changing. Make the regular tests for leaky or shorted transistors.

The normal drop between the collector and emitter of the (upper) regulator transistor is 15 V; if this is too low or if there's no drop at all (output voltage is too high), the transistor could be shorted. Because these are large, bolted-in transistors, they're not hard to remove for testing. When you return them, however, be sure to get the insulating washers, etc., in the right places.

# Transistorized dc-to-dc converters

Rather than having circuits where the ac supplies the power for the transistors, let's discuss how to make transistors convert dc to ac and then change the ac back to dc at a higher voltage. This is a circuit you won't see often. It's used mostly in middle-aged two-way FM transmitter power supplies and in auto radios built during the transitional period before they went all-transistor.

In the original auto-radio power supply, a vibrator interrupted the dc in the primary of a transformer so that you could step up the voltage to the 200 V or so needed for the plates of the tubes. The transistor version does the same thing; it supplies an ac voltage for the primary. You can step it up, rectify it, and come out with a high dc voltage. Transistors produce ac by oscillating. The ac in an oscillator circuit is a good substitute for the interrupted dc from a vibrator (it's also quieter).

These circuits are usually called *dc-to-dc converters* when the output is a high-voltage dc. However, there is another popular application; leave off the rectifier and filters and you have ac out. By making the output 117 V at 60 Hz, you can use ac equipment in cars, etc. Such dc-to-ac circuits are usually called *inverters* or *dc/ac converters*.

Figure 2-19 shows a typical dc-to-dc circuit with a common collector connection. This connection could be a transmitter power supply, a supply for a tube-type PA system, etc. The same principle is used in all such circuits. A transistor or pair of transistors is connected to the primary of the transformer. Feedback windings make the transistors oscillate. (This is a blocking oscillator circuit, but other types can be used.)

These oscillators often are run at frequencies above 60 Hz. The higher the frequency, the smaller the transformer can be. Power-line transformers must work on 60 Hz ac; vibrators run at about 115 Hz. You can make transistor circuits run wherever you want, though. Motorola, in some of its two-way radio power supplies, uses a frequency of about 400 Hz; others have used frequencies up into the thousands of Hz. With high-operating frequency, filter capacitors also can be much smaller for the same filtering efficiency. These characteristics make this circuit ideal for use in mobile and airborne equipment.

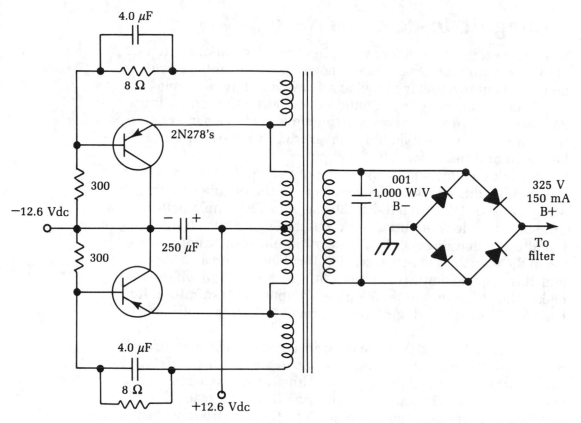

**Fig. 2-19**  *A typical dc-to-dc converter circuit.*

The output of the oscillator is a square wave, which is highly efficient and fairly easy to filter into dc. Bridge rectifiers are commonly used with standard pi-type filter circuits. Another dc-to-dc converter circuit is shown in Fig. 2-20.

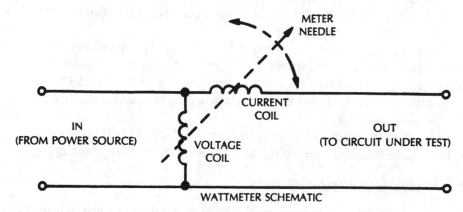

**Fig. 2-20**  *An alternate dc-to-dc converter circuit is shown here.*

# Testing dc-to-dc converters

In actual use, the dc-to-dc converter has been a remarkably trouble-free circuit. Some have been known to run for five years or more without ever having problems. However, they're simple to test. The dc output voltage should be measured first under full normal load, as in all other cases. If the converter is used in a radio transmitter, the voltage should be measured with the transmitter keyed on and tuned for full RF output.

If the dc is low, check the dc input voltage first. In mobile units, make this check in the vehicle with the engine running. Many cases of false trouble have been traced to simple battery drop-off after a long test under power. If the engine is running at a fast idle, the charging system will keep the voltage up to normal as when the vehicle is in actual service. Incidentally, on a 12 V system, this voltage should read almost 14 V. Running the engine can make a big difference in the RF power output of a transmitter. It has been known to bring RF power from 35 W to the full-rated 50 W output.

The ac voltage output from the transformer can be measured if you suspect the rectifiers of being weak. However, since this is a square wave, your rectifier-type ac voltmeter will not read the true value because it's calibrated on a sine wave. If it reads about 300 Vac, that's probably good enough. If so, then the trouble must be rectifiers or filters because the transistors are definitely in oscillation.

If the transistors are not oscillating, check all parts (since there are only five or six, it won't take long). If either of the 4 $\mu$F capacitors in the feedback circuits is open, the oscillator will not work at all, or not at the right frequency. If it is trying to run, there will be some ac in the secondary, but not at the right frequency, so the output will be very low.

The output waveform should be in a pretty good square wave, and it must be balanced. If it shows a decided imbalance, one of the transistors is not conducting as heavily as it should. Possible causes are leaky capacitors, leakage in the transistor itself, or an off-value resistor.

The circuits shown use pnp transistors. Npn transistors will do the same job, but the voltages will be reversed in polarity.

# IC voltage regulators

More and more pieces of modern electronics equipment are using an IC voltage regulator in their power supplies, especially any

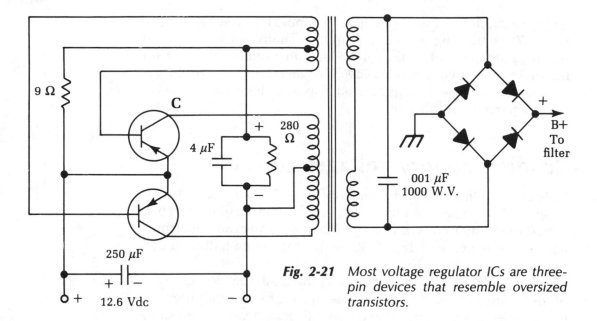

**Fig. 2-21**  *Most voltage regulator ICs are three-pin devices that resemble oversized transistors.*

equipment with digital circuitry, which requires very precise supply voltages and is very sensitive to power-line spikes.

The internal circuitry of voltage-regulator ICs is not dissimilar to the transistor voltage regulators discussed earlier in this chapter. Most voltage regulator ICs are three-pin devices, which resemble somewhat oversized transistors, as shown in Fig. 2-21. The unregulated input voltage is fed in to pin 1, and the regulated output voltage is taken off from pin 3. The middle pin is common to both the input and output circuits. In most cases, the common pin will be connected to ground, although there are some exceptions. The pin order varies among devices, so be sure to check the manufacturer's data sheet or the schematic.

If you suspect trouble with the voltage regulator, disconnect its output from the circuit being powered. Now measure the regulator's output voltage. If this voltage is not correct, the odds are strong that the voltage regulator is defective and should be replaced. To be positive, disconnect the unregulated input line from the regulator and measure this input voltage. It should be somewhat above the desired output voltage, but there might be considerable variation, depending on the specific design. As long as this input voltage is present and "in the ballpark," it is probably good and the voltage regulator is the culprit.

To test how well the unit is regulating the voltage, plug the equipment under test into a Variac. Vary the line voltage from about 90 V to 125 Vac to cover the normal range of typical line voltage variation the voltage regulator should be expected to cope with. While varying the Variac's output, monitor the voltage at the regulator output. This voltage should remain constant (or fluctuate over a very small range). Voltage-regulator problems can cause a variety of symptoms, such as insufficient or excessive gain or parasitic oscillations.

# Reading battery voltages correctly

This chapter has dealt with reading dc voltages in ac-powered supplies. However, don't forget the original source of power: batteries. A tremendous amount of battery-powered electronic equipment is in use today, so know how to check batteries correctly.

There's only one accurate way to read the voltage of a battery — under full load. This is easy; just turn the equipment on before taking a reading! Practically all batteries recover some voltage when the load is taken off. So if you want to know what voltage is present under actual operating conditions, check it with the equipment on and the batteries under load.

Even a dead dry-cell battery will read almost full normal voltage if you check it with a dc VTVM, which places virtually no load on the battery at all. When current is drawn from such a battery, the voltage drops to practically nothing. A brand-new standard dry cell reads about 1.64 V. This drops to about 1.4 V after an hour or so of use and gradually drops lower as the active materials of the battery are used up. When a cell reaches a load voltage of 1.1, it is considered "dead." Early battery radios were designed for a cutoff of 1.1 V per cell; transistor radios might work a little past this point, but the volume will be fairly low.

Incidentally, dry batteries are figured at 1.5 V per cell; so, a 9 V battery would be made up of six cells at 1.5 V apiece. Therefore, cutoff voltage for a 9 V battery would be 6.6 V under full load and so on.

Special battery testers are available. Such a tester is nothing more than a dc voltmeter combined with a shunt resistor to draw current from the battery being tested. A battery tester might be useful at times, but it is certainly simple enough to just turn the equipment on and read the battery voltage with an ordinary voltmeter.

In auto radios and other equipment used in cars, the power comes from the car's storage battery. Older cars have 6 V systems, and most new cars have 12 V systems. The 6 V type actually reads 6.3 V if the battery is fully charged, and the 12 V system, 12.6 V.

If the engine is running, the generator or alternator should be feeding current into the battery to keep it charged; the system voltage will, therefore, go up. The upper voltage limit is controlled by the car's voltage regulator. With the engine running at a fast idle, the voltage shouldn't go above about 14 V or so. If the voltage regulator isn't set properly, the voltage can go higher and cause damage to transistors, especially the high-power output types used in car radios now. If the system checks higher than 14 V with the engine running, have the voltage regulator adjusted by a competent mechanic.

The usual vacuum-tube voltmeter uses an input voltage divider whose total resistance is 11 MΩ. Some recent instruments have up to 16 MΩ input impedance. This resistance remains the same for every voltage range the meter may have.

Transistorized VTVMs have similar input resistance ratings. Field-effect transistors (FETs) often are used in the input circuits because of their high input impedances and because they are true voltage-amplifying devices. Ordinary bipolar transistors are current amplifiers.

The input resistance of the voltmeter might or might not be of significance, depending on the specific measurement being made. In power supplies, batteries, and other low-impedance, high-current circuits, the meter impedance — whether high or low — will not affect the circuit under test. If the voltage to be read has an ample current reserve, all meters will read the same.

On the other hand, in very high-impedance circuits — such as the plate circuit of a tube where the plate load resistance may be up to, say, 1 MΩ — meter resistance can make a lot of difference. Let's take a tube with a rated plate voltage of 50 V, which is supplied from a 150 V source through a 1 MΩ resistor. Ohm's law says that this means a normal plate current of .0001 amp, or 100 microamps (μA). If you try to read the plate voltage with a 1,000 Ω per-volt meter on a 50 V scale, total meter resistance will be 50,000 Ω. Placing this low resistance across the tube results in a shunting effect; current through the plate load resistor will increase because it now has a 50,000 Ω path from plate to ground. Since it doesn't take very much current through the 1 MΩ resistance between the plate and B+ to give a terrific voltage drop, the indicated plate voltage will probably be close to 0!

If you use a 20,000 Ω per-volt meter on a 50 V scale, the meter resistance will be 1 MΩ. Putting this in parallel with the tube, you'd read about three-fourths of the actual voltage, or about 37.5 V. This reading is still not accurate enough, unless you consider the voltage drop caused by the voltmeter.

However, if you hook a vacuum-tube voltmeter with an input impedance of 16 M$\Omega$ across the tube, it is more accurate. Now that the shunt resistance through the meter is much higher than the effective resistance of the tube, the meter reads much closer to the actual value. In fact, since most voltage readings in such circuits are now made with VTVMs at the factory, you would probably read 50 V.

When testing a circuit, you might find a little box in the lower corner of the schematic diagram showing which type of meter was used in taking the standard test-voltage readings. In some of the older sets, you'll find that a 20,000 $\Omega$ per-volt meter was used —in which case, if you use a VTVM, all voltages will seem high. So, to get an accurate voltage reading, you must always know the test conditions and the type of instrument used to make the standard readings.

Digital voltmeters normally are used in the same manner as VOMs, but in terms of their electrical characteristics they are closer to VTVMs. Digital voltmeters generally feature a very high input impedance —1M$\Omega$ per volt is typical. A digital voltmeter makes precise readings very easy because a numerical value is displayed directly and there is none of the inherent ambiguity involved in determining the exact position of a meter pointer on a scale.

On the other hand, in many cases you need to monitor a continuously changing value. You can monitor the trend of the change easily by watching the up or down movement of a meter pointer, but on a digital voltmeter you would only get a blur of meaningless numbers. The best type of voltmeter to use depends on the nature of the specific application at hand.

# 3
# Current tests

CURRENT MEASUREMENTS AREN'T AS SIMPLE TO MAKE AS VOLTAGE measurements. The meter must become part of the circuit instead of touching across it, which takes more time. Also, you have to be very careful—picking the wrong current range or shorting the load to ground accidentally can blow a meter very quickly. For this reason, many technicians are reluctant to use current tests. However, current measurements can give a great deal of information about a circuit in a short time. With reasonable care, there is no more risk of meter damage in this test than in any other.

You can make the job of getting the meter into the circuit much easier with some special equipment. Test adapters of various types allow current measurement by plug-in testing without unsoldering wires. Some types of adapters can be bought ready-built, but you might have to make a few. They're well worth the little time it takes in terms of bench time saved.

Reading the input current drawn by any electrical apparatus can tell you the total wattage being consumed; simply multiply the current by the applied voltage. The actual wattage consumed is a valuable piece of information because the rated wattage is almost always given in the service data for the apparatus. If a device is taking more power than it should, there is definitely something wrong—a leakage, a short circuit in the power supply, etc.) By measuring input current you can, for example, check a power transformer for an internal short.

The input current test works on ac and dc equipment, all the way from a tiny transistor radio to a 5,000 W transmitter. Current measurements within a circuit are essential in transmitter testing and in all kinds of high-power work such as PA systems and high-power amplifiers. In high-power transistor amplifiers, a current measurement can tell you if transistors are leaky, indicate whether bias voltages are correct, and give you other useful information.

## Power measurements

If a TV set, amplifier, etc., becomes too hot as it plays, you need to know if it is actually drawing too much power, and if so, why. The input wattage measurement of any piece of ac-powered equipment is a very good clue to what's going on. You can check the actual power being consumed to pin down a short or leakage in the power supply circuits, or even an open circuit—one that isn't drawing as much power as it should.

The watt is a unit of power that is always a product—volts multiplied by amps. So an ordinary voltmeter or ammeter will not read power—you have to take simultaneous voltage and current readings and then do the math. Of course, there are wattmeters that will do this for you. They're actually combination volt-ammeters, having a current coil in series with the line and a voltage coil across it. Both coils affect the position of the meter needle: the amount of deflection depends on the product of the currents in the two coils, so the scale can be calibrated directly in watts (see Fig. 3-1)

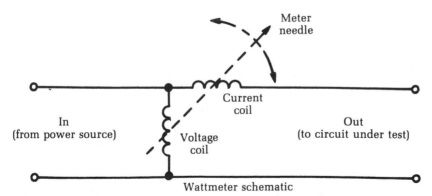

**Fig. 3-1**   *The current through the two coils affects the amount of pointer deflection.*

There are many uses for this instrument. However, wattmeters are not commonly found in service shops because they are fairly expensive. Some shortcut tests can give the same information but use more common test equipment. Any of the wattage tests discussed can be made accurately with the three substitute testers described later.

## Measuring the dc drain of an auto radio

Measuring the current drain of an auto radio is probably the simplest current test. All you need is a 0-to-10-amp dc ammeter, connected to one of the power-supply leads to the auto radio, as shown in Fig. 3-2. The rated current drain for the specific set should be in the service data.

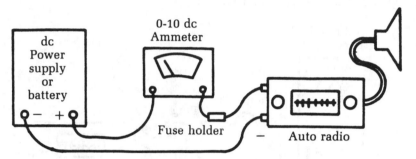

**Fig. 3-2**    An ammeter can be connected to one of the power-supply leads of an auto radio.

Transistor equipment generally draws a relatively low amount of current, usually under 1 amp at 12 V. Older tube sets with vibrators drew considerably more current. Drains as high as 8 to 12 amps were uncommon. Some early hybrid models that used tubes for the early stages and transistors in the output stage had current ratings in the 1- to 1.5-amp range.

If you use an ac-powered bench supply for operating auto radios for test, the supply will probably have a dc voltmeter and an ammeter built in. If not, you probably can use your VOM for the ammeter; such meters usually have a dc range of 10 amps or so, which is ample for most auto radios. Be sure that the supply voltage is set to the state level because this affects the current drawn and the wattage.

If the tone of an auto radio is not as good as it should be and the current is either more or less than the rated value, check the bias on the output transistor(s). The output stage causes the heaviest current drain of the whole set.

# Dc current measurements in transistor portable radios

Current measurements are invaluable in servicing small transistor radios, particularly the subminiature types. Because of their small size, it's hard to get into these circuits. So, we take current readings, which we can do from the outside, and get all the information possible before we start taking things apart.

Practically all the bench power supplies used for testing these radios have a dc milliammeter as well as a dc voltmeter. If yours doesn't have one, you can always use the 0- to 25- or 0- to 50-mA range of the VOM, as in Fig. 3-3.

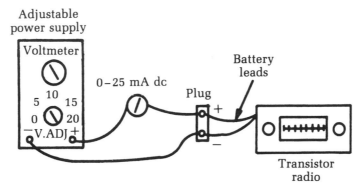

**Fig. 3-3**   *A VOM can be used to monitor the current supplied by a bench power supply.*

A set of connector harnesses can speed up the work. Take the battery terminals from a dead battery and solder a set of test leads to them, watching polarity. Now, when working on a radio using this size battery, you simply snap on the connector. It's handy to have a harness for each of the common sizes of batteries.

A set of test leads with alligator clips is very useful. For example, you can use them to connect a milliammeter in series with the radio battery by turning the battery plug sidewise and clipping on, as in Fig. 3-4.

If the radio uses penlight batteries in holders, you can get your meter into the circuit with an adapter like the one illustrated in Fig. 3-5. Get a strip of heavy insulating paper (such as the fish paper used in electrical work) about ½ inch wide and 2 to 3 inches long. Cement a thin strip of brass (shimshock) to each side, and solder test leads to the ends of the strips. Be sure the brass strips are cut a little smaller than the insulator. To use this adapter, in-

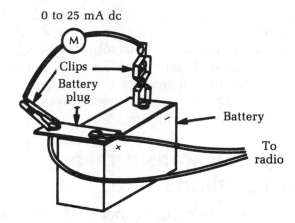

**Fig. 3-4**  *Test leads with alligator clips can be very useful.*

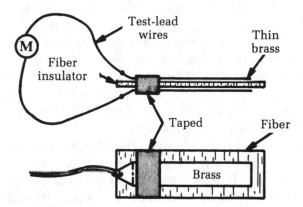

**Fig. 3-5**  *This adapter can be employed to use a meter with a circuit pow-ered by penlight batteries.*

sert it between any two batteries in the string and connect the test leads to the milliammeter. It's easy to get the adapter in place if you lift the ends of the two of the batteries, put the adapter between them, and then push them back into the holders.

Check the service data to determine what the current drain should be. In a typical six-transistor portable, it might run less than 10 mA at minimum volume and about 15 mA to 20 mA at full volume. Maximum current depends on how much audio power output the set has.

There are many uses for the input current test in addition to checking bias, battery life, etc. You can even use it as an alignment indicator. The total current drain of a transistor radio depends on the audio output; if you use an audio-modulated signal for alignment work, the current drain will be directly proportional to the amount of audio signal. So feed in an IF or RF signal and tune for maximum current. The volume control can be set to give a convenient amount of current. This setting and that of the RF signal generator output shouldn't be changed during alignment unless the signal becomes too high and threatens to overload the receiver or cause clipping.

# Testing power transformers for internal shorts

Many sets have "hot" power transformers. They might smell bad, have tar running out, etc., and in general show all the symptoms of being hopelessly burned up. But are they? The big question is always this: "Is the transformer broken down internally, or is the overheating a result of a short in the load circuits?" In all cases, you must know if the transformer itself is bad before you can make any estimate on the job. So, check it first.

Remove the rectifier tube or disconnect the silicon rectifiers, etc. In a few sets, you can do this by pulling the V+ fuse. In any case, be sure that the rectifiers are completely disconnected from the power-transformer secondary.

In a tube set, you should open the filament circuit by disconnecting one wire from the power transformer. (You could pull all the tubes, but that takes longer!) If the filament circuit is center-tapped, you'll have to open both supply wires. In all tests, make sure that there is no load on the power transformer.

Plug the primary of the power transformer into the wattmeter and turn it on. If the transformer is not internally shorted, you'll see a very small kick of the meter needle as the magnetic fields build up, and then the reading will fall back to almost zero. This reading, usually 2 W to 3 W at most, is the "iron loss" and is normal. If there is a shorted turn anywhere in the transformer, however, the wattmeter needle will come up to 25 W or 30 W. If the short is in one of the high-current windings, you'll see a full-scale, needle-slamming reading, which means that the transformer is *definitely* bad. Recheck to make sure that all loads have been disconnected. (Even two pilot lights can show a reading of about 5W).

There is also a no-instruments-at-all test you can do on a power transformer. Hook up the transformer with no load on it, and leave it on for 5 to 10 minutes. If the transformer gets too hot, it's bad. A good transformer will get just barely warm running no-load. A badly shorted one will heat up and smoke.

## Substitute power testers

Let's discuss the kind of equipment you can substitute for a watt-meter to get the same results. Since you need a "volts times amps" reading, you have to do your own arithmetic. The ac line voltage is usually specified as being 117 V rms, but let's use 120 V for simplicity. If you know the line voltage, measure the current and get watts by multiplying. For example, if the rating plate of the equipment under test reads 240 W your current should measure 2 amps.

You can use a 0-to 5-amp ac ammeter (Fig. 3-6) of the panel-meter type for such measurements. Also, an adapter is available that makes a clamp-on ammeter out of a pocket VOM. It has a current transformer made so that the core can be opened up and clamped around either one of the ac wires. When the adapter is in place, the meter reads ac current on the ac volts scale.

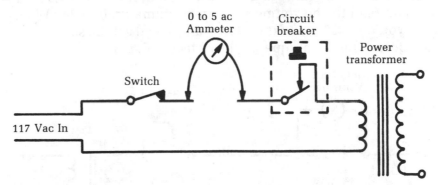

**Fig. 3-6**    *A 0 to 5-amp ac meter in series with the input can be used to measure wattage.*

If you have one of these adapters (Fig. 3-7), be sure to place the clamp-on core around only one of the wires; if both wires get inside it, you won't get any reading. On TV sets, you can usually get at one wire at the ac interlock, at a wire going to the switch, etc. You can now read current, multiply it by line voltage, and end up with the wattage.

However, ac ammeters aren't common in service shops either. Let's figure out another way to measure power. Use the ac voltmeter from the VOM. Get a fairly accurate 1 Ω resistor; the standard 5 W or 10 W wirewound types are an "automatic" Ohm's

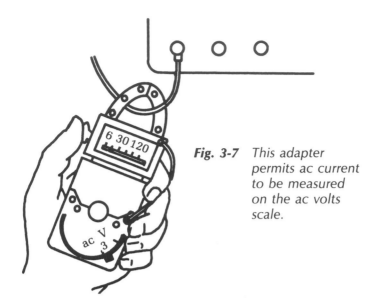

**Fig. 3-7** *This adapter permits ac current to be measured on the ac volts scale.*

law computer. Using $E + IR$, Ohm said that for every ampere of current that flows through a 1Ω resistor, it drops 1 V across it. So, you can read the ac voltmeter as an ac ammeter. In a 240 W circuit, you'd read 2V (and therefore 2 amps) and get the same result as before. This technique is illustrated in Fig. 3-8.

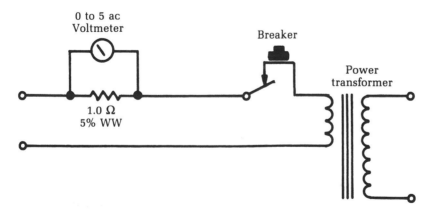

**Fig. 3-8** *Ohm's law can be used to measure ac current on the ac volts scale.*

The method just described is a valuable test for sets using small circuit breakers in the primary of the power transformer. If the breaker kicks out at odd intervals, check to see whether it is caused by an intermittent overload in the secondary circuits or to an intermittent breaker. The kickout value of current is always specified on the breaker itself. For example, a typical unit might be rated "hold 2.2 amps; open at 2.5 amps." If you read the actual current in the circuit and find that the breaker is kicking out at 2.2 amps, replace the breaker and that usually fixes the trouble!

# Finding overloads in power supplies

After clearing the power transformer from suspicion in a fuse-blowing or overheating problem, you can put the loads back on one at a time and find out which one is faulty. One good test is to leave the filaments open and hook up only the V+. Check input power again. If it shows more than 30W to 35W, look out! If the current drain is as much as 75W to 100W with only the B+ circuits hooked up, there is definitely a short circuit in one of the branches.

Leaving the filaments off will raise the V+ voltage by taking off the normal loading. This test also can help to break down leaky parts and give you a nice, definite indication, such as a thin pillar of smoke. So do this test with care and with one hand on the switch!

If the transformer and V+ filter circuits check okay, you can read the V+ current drain by hooking a 0 to 500 dc milliammeter into the V+ filter output, as in Fig. 3-9. The normal current drain will always be given on the schematic; for example, 350 V, 260 mA — and so on. If you get something like 300 V at 290 mA, look out. Something is drawing more than its normal current.

The excessive current can be traced by voltage readings in individual circuits, especially if they have their own individual dropping resistors. Look for the resistor with the greatest percentage of drop. For instance, if you have taps for 350 V, 250 V, and

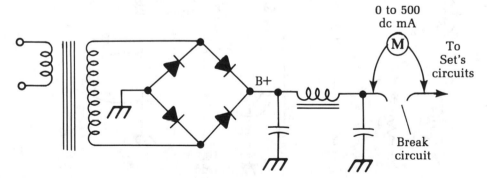

**Fig. 3-9**   *A 0 to 500 mA milliammeter can be used to detect overloads in the V+ filter output.*

150 V coming off the V+ filter output, and the 350 V and 250 V lines read about 10 V low while the 150 V line reads about 50 V maximum, the trouble is in the 150 V line. For a quick check, disconnect the 150 V line and see whether the others jump back up to normal (or probably a little above because of the reduced loading). Having traced the trouble to a single circuit, you can chase down the short with an ohmmeter.

# Setting the bias of a power transistor with a current meter

Transistor bias voltages are crucial. A change of only a small fraction of a volt can cause a transistor to cut off completely or draw a very high current—often enough to overheat the junction and destroy the transistor. If you suspect that a power transistor is drawing too much current, the only way to check it is to insert an ammeter and read the current directly.

Figure 3-10 shows this situation on a typical single-transistor output stage. The bias is adjusted by means of a series resistor in the base return circuit. Open the bottom of the output transformer primary circuit, as shown, and insert a 0- to 1-amp dc meter. With this connection, read the collector current; its value is determined by the base bias. (You can consider this as either a

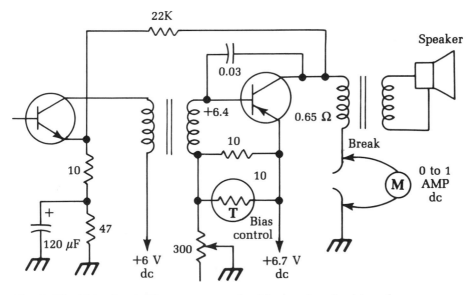

**Fig. 3-10**   *This circuit illustrates a method for setting the bias of a power transistor with a current meter.*

voltage bias or a current bias; if you change current, you change voltage and vice versa, so don't be confused about it.)

The circuit shown happens to be from a 6.3 V radio, and the power output is not very high. The rated values for this circuit are as shown, and the bias control should be adjusted to give a collector current of 550 mA to 600 mA (0.55 to 0.6 amp).

The collector goes directly to the ground through the primary of the output transformer. This winding has a very low dc resistance, so a dc voltage measurement here would be hard to read unless you had a meter that read millivolts. This explains why you take a current reading, which has more readable values. The actual current will be different on other sets, but the principle is the same.

For a given auto radio, details on how to adjust the output current are in the service data. Check this data before you make the test because auto radios differ in current values. The current in a given circuit depends on the type of transistor used and the way the circuit was designed. You need the exact value so you can adjust for correct current in the set you are working on. The bias value sets the operating point of the transistor.

In two-transistor push-pull or transformerless output circuits, correct collector current is especially important. Most of these circuits work in class B: one transistor amplifies the negative half of each cycle; the other, the positive. If the bias isn't exactly right, you get what is called *crossover distortion* at the point where one transistor stops conducting and the other starts. You can see this distortion with a scope; there will be a decided break or notch in the signal waveform where it crosses the zero line. Some sets use a combination bias—the output stage actually works in class A on small signals, shifting to class B on large ones. Bias is very important on these circuits. Combination-bias circuits are also found in hi-fi stereo amplifiers and are adjusted in exactly the same way as car-radio circuits.

## Measuring large currents

For most modern electronics work, you will be dealing with fairly small currents, usually in the milliamp range. (One milliamp equals 0.001 ampere.) Rarely will you encounter currents of more than 1 or 2 amps, but there are exceptions in some types of equipment.

Most multimeters aren't designed to handle relatively large current measurements. The highest current range on many multimeters is only about 500 mA (0.5 amp), especially on less expensive models. But what if you need to measure a large current value, say 12 amps? You'll only peg the meter's needle at the high end of the scale. You won't get a useful reading, and you might possibly damage your multimeter.

You can increase the range of any milliammeter (or ammeter) by adding a shunt resistor in parallel with it, as illustrated in Fig. 3-11. This trick works because of Ohm's law, which states that:

$$I = E/R$$

where $I$ is the current in amperes, $E$ is the voltage in volts, and $R$ is the resistance in ohms.

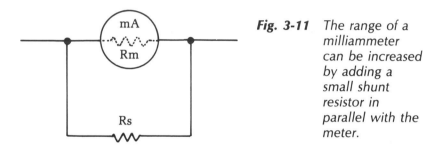

**Fig. 3-11**   *The range of a milliammeter can be increased by adding a small shunt resistor in parallel with the meter.*

The total effective resistance in this case is equal to:

$$R_t = \frac{R_m \times R_s}{R_m + R_s}$$

where $R_m$ is the internal resistance of the milliammeter, and $R_s$ is the added shunt resistance. The total effective resistance ($R_t$) is always less than either of the component resistances in the combination ($R_m$ or $R_s$).

As an example, let's assume the internal resistance of the milliammeter is 30 $\Omega$, and a full-scale reading is obtained with a current of 0.5 amp (500 mA). This will occur when the voltage source equals:

$$
\begin{aligned}
E &= IR \\
&= 0.5 \times 30 \\
&= 15 \text{ V}
\end{aligned}
$$

We want to extend the range of the meter to 5 amps (5,000 mA). The meter itself can't handle more than 500 mA, so the remaining 4.5 amps must be carried by the shunt resistor. What should its resistance value be?

In a parallel combination, the same voltage will pass through both individual resistors. We already know that a supply voltage of 15 V gives the maximum current through the meter. We know the supply voltage (15 V) and the desired current (4,500 mA) to be carried by this resistor. All we have to do is to rearrange Ohm's Law to solve for the unknown resistance:

$$R = \frac{E}{I}$$
$$= \frac{15}{4.5}$$
$$= 3.3333 \ \Omega$$

Using a 3.3 $\Omega$ shunt resistor will increase the milliammeter's range by a factor of ten. Notice that the value of the shunt resistor is typically very small. The greater the desired increase in the meter's range, the smaller this resistance must be.

This is all well and good, but it assumes you know the internal resistance of the original, unmodified milliammeter. What if this information is not readily available? You can't measure this resistance directly with an ohmmeter.

A solution to this problem can be found with the test rig circuit illustrated in Fig. 3-12. This test circuit will work fine for either a bare milliammeter or the milliammeter section of a multimeter. Use 10K (10,000 $\Omega$) potentiometers for R1 and R2. Resistor R3 is simply a protective current-limiting resistor, and its exact value is not terribly crucial. Use a 4.7K (4,700 $\Omega$) or 3.9K (3,900 $\Omega$) resistor, or anything you happen to have available in this approximate range.

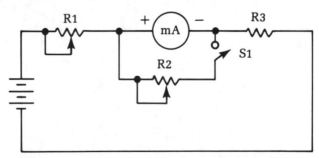

**Fig. 3-12**  *This test circuit can be used to determine the internal resistance of a milliammeter.*

Start out with potentiometer R1 set for its maximum resistance (lowest current flow) and switch S1 open. At this point potentiometer R2 is not actually part of the circuit. Slowly decrease the resistance of potentiometer R1 to increase the current flow through the circuit, and therefore the reading on the milliammeter. Keep going until the meter shows its full-scale reading. Now, leave potentiometer R1 set as it is. Do not move it.

Close switch S1 to add potentiometer R2 to the circuit in parallel with the meter's internal resistance. Adjust potentiometer R2 until the meter gives an exact half-scale reading. At this point, the two parallel resistances (R2 and the meter's internal resistance) must be equal. Being careful not to change the setting of potentiometer R2, open switch S1 and use an ohmmeter to measure its resistance. The milliammeter's internal resistance has the same value as you read across this potentiometer.

# 4
# VOM and
# VTVM tests

THIS CHAPTER DESCRIBES A WIDE RANGE OF APPLICATIONS FOR THE technician's right arm: the multimeter.

## Measuring contact resistance with a VOM

Once in a while you need to test a switch or relay contact to see if it is making good contact. A bad contact is annoying, especially if it is intermittent. However, there is one positive test: checking the resistance of the contact by reading the voltage across it in actual use. This test hook-up is simple, as illustrated in Fig. 4-1.

The contact resistance of a good switch should be 0 — there should be absolutely no voltage drop across the contacts under full normal load. If the contact surfaces become dirty or burned, there will be resistance. Where there's resistance, there's heat, voltage drop, etc.

To check for contact resistance, set your voltmeter to the full line voltage of the circuit and connect it directly across the contacts. The voltage can be either ac or dc; just set the meter to read the full voltage that the switch is breaking. With a sensitive meter, you'll see the full voltage across an open switch, even if the load circuits have fairly high resistance. Now, close the switch; the voltage should drop to 0. Turn the switch on and off several times to see if the voltage does drop to 0 every time.

If there is any contact resistance, you'll see a small voltage reading when the contacts are closed. In most cases, this can be

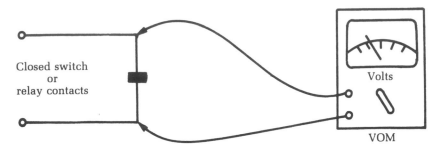

**Fig. 4-1** *It is very simple to measure contact resistance with a VOM.*

read with your initial meter setting. If you want to, turn the meter to a lower voltage scale for a more accurate check, but be careful. The contact could be intermittent, and any resulting surge of voltage could damage the voltmeter (unless it is a VTVM).

With the meter set on the maximum voltage scale, tap the switch with the handle of a screwdriver to see if you get any difference in the reading. It's a good idea to do this before going to a lower voltage scale with a VOM.

It takes only a very small voltage to show you that there is unwanted contact resistance. Even 0.5 V means that there is some resistance between the contacts. This resistance will tend to increase in time, and the switch might eventually burn out. (Resistance makes heat; heat makes more resistance—it's a vicious circle.)

If the contacts are accessible as in some relays, you should clean them with very fine sandpaper or crocus cloth and a contact-burnishing tool, or even a piece of cardboard. After cleaning, check again for zero resistance and zero voltage drop.

# Measuring voltage drop in long battery wires

There might be some mystery when troubleshooting too much voltage drop in the supply wires. Take a typical case: a two-way radio transceiver is mounted in the trunk of a car and connected to the car's battery by long wires. If these wires aren't heavy enough or if there is contact resistance in a switch or relay, the radio supply will be low, and you'll have a "mysteriously" weak set.

The first step in such a case is to read the battery voltage at the radio power-supply plug. If the voltage is below normal, then go back to the battery itself and read its voltage. The radio must be turned on so you get a full-load reading. In most cases, it's desirable to key the transmitter to place maximum load on the power supply.

If the battery voltage is normal and there's a loss between the battery and the radio, the next step is to find out just what part of the supply circuit is guilty. Do so by taking more voltage readings — this time across the wires themselves. You'll probably need an extension lead for one of the voltmeter test leads because you have to take a reading between two points that are 18 to 20 feet apart! Don't worry about the length of the test lead. Because of the small current drawn by the voltmeter, voltage drop in the test leads will not affect the reading much.

Connect one test lead to the ungrounded post of the battery itself. If this post is positive, as it is in most American cars, put the positive meter lead here. Make sure that you have a good, clean connection; a clip will be needed to hold the lead in place. Now put the other test lead on the positive (or "hot") wire of the radio power-supply plug, as at A in Fig. 4-2. Turn the radio on; there should be no voltage reading on the meter. Normally, you'll be able to see any significant voltage drop, since it must be from 1 V to 3 V to cause trouble.

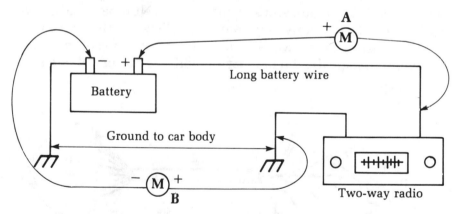

**Fig. 4-2**  *Special techniques are used to measure the voltage drop in long battery wires.*

If you see no voltage drop on the positive lead, put the negative lead of the voltmeter on the grounded battery post. Then take another reading, this time to the ground point in the car's trunk or at the ground lead of the power supply plug (B). Again, you should see no voltage drop. If, however, the ground path is not perfect — for example if there are painted joints between this point and the battery ground — you will see some voltage.

By doing these two tests, you can pin down the cause of any excessive voltage drop in the supply circuits. Don't forget that these wires also go through switches and relays. To get away from possible voltage drops in automotive switches, most two-way radios and similar high-current equipment have relays that do the actual switching on or off; the car's ignition switch is used only to control this relay. Make sure the supply circuit does not go through the ignition switch itself; such switches are too light for this purpose.

# Identifying uncoded wires in multiconductor cables

Identifying the wires in a large group of non-color-coded wires is difficult at best. You find this problem in intercommunication systems, where someone has bunched assorted wires without color-coding or identifying them. What do you do?

Get a 9 V battery and a connection strip. Solder a pair of short test leads to the strip, with alligator clips at each end of the leads. You can hook a milliammeter in series with one lead, or use a pilot light (a 12 V type). This is shown in Fig. 4-3. Now all you need is a 0 V to 10 V dc voltmeter and a set of wire tags. Numbered adhesive strips are sold by many electronics suppliers for this purpose.

The testing takes two people. Find the two ends of the cable you want to identify. Put one person with the battery at one end,

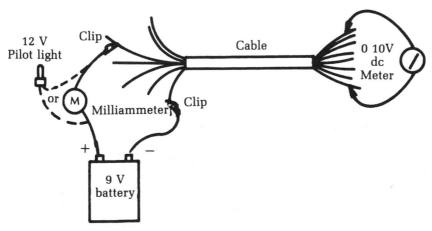

**Fig. 4-3**  *A handy method for identifying uncoded wires in a multiconductor cable.*

connect the battery to any two wires in the cable, and mark the negative wire NO. 1 and the positive wire NO. 2. At the other end of the cable, with your voltmeter, start checking between all possible pairs of wires for voltage. This is the hard part because you have so many possible combinations. If the wires have continuity, you'll eventually find the right pair. Mark the negative wire NO. 1 and the positive wire NO. 2. To signal the person at the other end, short the two wires together a couple of times. If he's using the pilot light, it will flash, indicating that you've found the first pair.

Your partner leaves the negative clip connected to the same wire (NO. 1) and moves the positive clip to any one of the others. You leave your negative clip attached and search with the positive one until you find voltage on another wire. Mark this wire NO. 3; then short it to NO. 1 to flash the signal light. Your partner should have marked the positive wire NO. 3 while you were searching and now moves the clip.

This technique also serves as a good short or leakage test. Even after you locate the wire with voltage on it, check all the wires. See if you get a voltage reading on any of the others. You shouldn't, unless there is water in the conduit, an accidental short, etc. If there is a short, this method will reveal it quickly. Your partner's light will turn on the instant it's connected to the shorted pair.

# Testing capacitor leakage with a VTVM

In checking small capacitors, find out if they are open, leaking, or shorted. Because of the way such capacitors are built, it's almost impossible for them to change value. A 0.001 $\mu$F capacitor is likely to remain a 0.001 $\mu$F no matter what happens to it. Rather than check the value as with electrolytics, do a condition test to determine whether the capacitor is functioning. A leakage of as little as 200 to 300 M in a high-impedance circuit can upset things considerably, so it's important to test for capacitor leakage when trouble occurs in these circuits.

A capacitor tester that has an insulation-resistance function is handy. All you need is a VTVM. Figure 4-4 shows how to do it. Disconnect the capacitor and hook one end to the dc-volts probe of the VTVM. Now touch the other end to any convenient source of V+ voltage. The actual voltage isn't too important, as long as it is about the same or somewhat higher than the voltage applied to the capacitor in the circuit. For large power-supply capacitors, especially in older tube equipment, about 200 V to 250 V is fine.

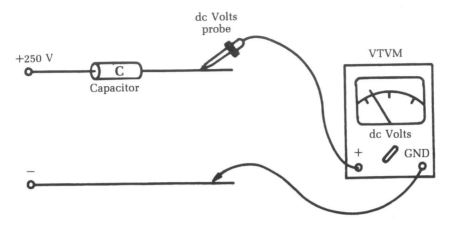

**Fig. 4-4**  *A VTVM can be used to test capacitor leakage.*

In modern solid-state (especially IC) circuits, much smaller capacitors are normally used, with rated working voltages of 25 V or even less. Do not exceed the maximum rating of the capacitor under test or you could destroy it. For small, low-power capacitors, the ohmmeter battery in a VOM is sufficient. On the other hand, too low a test voltage might not give a readable result.

You'll see an initial kick of the voltmeter, caused by the charging current of the capacitor. If there is little leakage, this initial deflection gradually will ease off and the meter needle will go back to zero. The larger the capacitor value, the longer this takes. If you're in a hurry, you can touch the VTVM probe tip and ground with your fingers to discharge the capacitor faster.

Now, watch the meter needle. If it never reaches zero, or if it starts back up again after you have brought it to zero by touching the probe to ground as described, there is leakage in the capacitor. You're making a voltage divider that consists of the input resistance of the VTVM and the capacitor. Leakage current gives a voltage drop across the meter—and a voltage reading.

The actual voltage isn't too important; the fact that there is any voltage at all is what you're looking for. In coupling-capacitor circuits, even a 1 V leakage can be too much. It causes a leakage of positive voltage from the plate of one stage onto the grid of the next stage, canceling out part of the negative bias voltage and perhaps throwing the tube onto the wrong part of the curve and causing a severe distortion.

With transistors in similar circuits, capacitor leakage is even more important. It takes only a tiny fraction of a volt to make a transistor cut off or run wild. The latter can cause overheating and avalanching, and can destroy a transistor quickly. For best results, check any new capacitor before putting it into circuit. Such a check can save a lot of time if the new capacitor does happen to have a little leakage! We have a tendency to automatically assume that a new part is good. This isn't true in all cases; check first, and you'll be sure.

# Measuring very high resistances

You might have occasion to check high-value resistors that are far beyond the range of your ohmmeter scales. There is a way to do this—in fact, there are several ways.

The easiest is with a resistor of the same value or one as close as possible. If you don't have an exact duplicate, you can connect several resistors together temporarily to get the necessary value. For example, for a bleeder resistor used in color TV focus-rectifier circuits—66 M—hook three 20 M resistors and a 6 M resistor in series; or use three 22 M resistors; etc.

For the test, connect the combination in series with the suspected resistor and connect the whole string across a source of voltage. The V+ voltage of the TV set is handy, but you can use any voltage, even the ac line if you want. Take a voltage reading across one of the resistors, as shown in Fig. 4-5, and note the reading. Next, check the voltage across the other resistor.

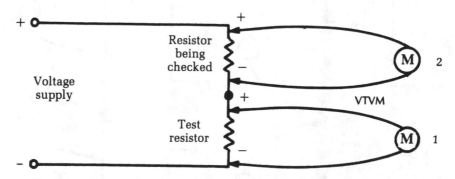

**Fig. 4-5**  *Very high resistances can be measured using this method.*

Because you have deliberately made a divide-by-two voltage divider, the two readings should be almost the same if the resistor under test is okay. If the difference is not more than 10 percent, the resistor is probably good. (The amount of difference you should tolerate depends on the resistors used; if they are 5 percent resistors, then cut down your tolerance to this amount, and so on.)

If you don't have the right resistor values, you can still use some that will make the total resistance of the voltage divider come out at a value that is easy to interpret. For example, with the 66 M resistor, you could use a single 22 M test resistor. Your whole circuit would then offer 88 M (66 + 22); your voltage reading across the 22 M resistor would be one-fourth of the total voltage. If you used 200 V, you'd read 50 V across the 22 M and 150 V across the 66 M.

Note that this method takes the meter resistance out of the picture only if all resistors are equal. If you had to figure the actual voltage present across each resistor, you'd have to figure out the shunting effect of the meter resistance.

## Checking diodes by the balance method

You can use the method described for checking very high resistances to also check diode back-resistance. You can check ordinary diodes, such as afc, video and sound detector, etc. with an ohmmeter, as will be shown in a later chapter. However, if you run into special units, such as the high-voltage diode rectifiers used in focus and boosted-boost circuits in color TV sets, the ohmmeter simply won't reach the resistance range needed.

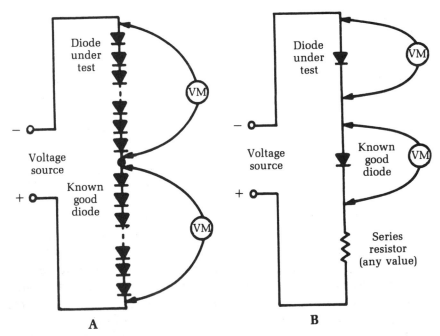

**Fig. 4-6** *Diodes can be checked with the balance method.*

The duplicate-and-balance method shown in Fig. 4-6A is the easiest. Connect the suspected unit in series with another one just like it, and feed a voltage to the combination for equal voltage drops across the two units.

You also can use this method for testing low-voltage diodes and for checking balance on afc diodes, ratio-detector diodes, and diodes used in any of the circuits of FM multiplex receivers and decoders. With these low-voltage units, connect the diodes in series and add a series resistor to keep the current within safe limits (see part B of Fig. 4-6). Now hook a small test voltage across the combination and check the diode voltage drops for balance.

As a general rule, if any diode — from the 5,000 V focus rectifiers down to one of the tiny video-detector diodes — is really bad, the defect will show up in the front-to-back resistance ratio, which can be measured by this test. You can feed an audio or RF signal across the combination and read the result with an oscilloscope or a good ac VTVM.

# Checking video-detector diodes with an ohmmeter

You can check the small video-detector or signal diodes very accurately with an ohmmeter. A good diode shows a small resistance in one direction and a very high one in the other. The actual resistance readings seem to depend on the battery voltage used in the ohmmeter. The forward resistance varies with voltage and diode types but will probably be in the neighborhood of 100 to 200 ohms. The reverse-bias measurement should read several hundred thousand ohms.

Practically all VTVMs use a low-voltage battery in the ohmmeter; the average voltage is about 3 V. Most VOMs use a 1.5 V battery on the low-resistance ranges, but use up to 22.5 V on the highest resistance scale in the megohm range. Diodes should be tested on low range — a scale of 0 to 2,000 $\Omega$ or 0 to 5,000 $\Omega$ — because such a range will use a low-voltage battery.

Find out which of your ohmmeter leads is positive; that is, which one is connected to the positive side of the ohmmeter battery. Check yours with another voltmeter and mark the positive lead. A voltmeter check also tells you exactly what voltages are used in your ohmmeter on various ranges.

If the positive ohmmeter lead is placed on the cathode of the diode and the negative on the anode, you get a high-resistance reading. Reverse the leads for a low-resistance reading. The ratio between the two should be very high for best performance in video detectors and similar circuits.

Actually you don't have to bother with polarity. With a high resistance one way and a low resistance with reversed leads, the diode is okay. However, with an unidentified diode or one so small that you can't tell what the markings are (not uncommon), use the ohmmeter to identify the elements. In the forward-bias direction, the positive lead is on the anode (triangle) and the negative is on the cathode (bar) of the diode. This test is illustrated in Fig. 4-7.

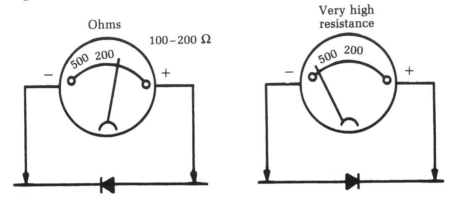

**Fig. 4-7**   *This is a technique for checking video-detector diodes with an ohmmeter.*

# Checking video-detector diodes in-circuit with an ohmmeter

Whenever you run into a white-screen symptom in a TV and the cause is not an agc block or a dead IF tube or transistor, it could be a bad video-detector diode. These diodes are usually contained in a small can, and they shouldn't be unsoldered and resoldered too many times. You can make a definite and reliable test without moving a wire: the ohmmeter will tell you from outside whether the diode is good or bad.

Figure 4-8 shows a typical video-detector circuit. The IF signal comes to the diode from a secondary winding on the last IF transformer; the other end of the winding is grounded. (This is a series-diode detector, but shunt detectors can be checked in the same way.) The coils are peaking chokes, which give the circuit a wideband response, and the video-diode load resistor is the 3,300 Ω unit connected from the diode output circuit to ground.

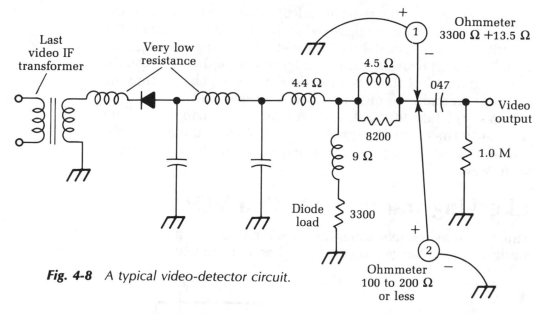

**Fig. 4-8**  *A typical video-detector circuit.*

To check the diode, first hook the negative lead of the ohm-meter to the video coupling capacitor, as shown. That way you can read the dc path through the peaking coils and diode to ground. The positive ohmmeter lead is connected to ground (1 in the figure). This is the reverse-bias direction, so you should read nothing but the load resistor plus the small resistance of the two peaking coils — actually, 3,318 Ω (but if you get 3,300 Ω, it's okay.)

That's one check. Now, reverse the leads — positive to the coupling capacitor, negative to ground (2 in the figure) — and read the diode in the forward-bias direction. Your path is through the low forward resistance of the diode with the 3,300 Ω resistor in parallel, so the measurement should read something like 150 Ω. Simply ensure there is a good, high ratio between the two read-ings.

If the reading is infinity on both tests (no reading at all), the diode or one of the peaking coils is open. It doesn't matter which; you're going to have to take the circuit apart to find out which one and fix it. Check each of the coils for continuity before you de-solder the diode; such coils sometimes open because of corrosion on the fine wires. If you have a coil shunted with a resistor — such as the last coil in the circuit shown, which is shunted by 8.2K — the value of the reading can tell you which coil to suspect. For ex-ample, if you read 8.2K + 3.3K from grid to ground (in the high-resistance direction) you know the last coil is open, so check it first.

If the reading is the same in both directions, and very low, the diode is shorted. You can figure what the correct resistance of the circuit should be by totaling the resistances of the peaking coils and diode load resistor, etc. If you have any doubts, you can lift the ground end of the diode load resistor, thus eliminating the shunt path, and read only the diode's resistance.

In shunt detector circuits, where the diode is connected from the transformer output to ground, you can trace the dc path for the ohmmeter reading and catch an open or shorted diode in the same way.

# Checking transistors with a VOM

The VOM makes a good transistor checker. It can be checked with an ohmmeter. Figure 4-9 shows the basis of the test.

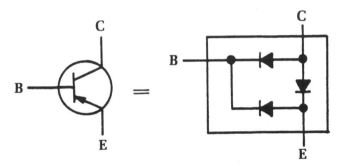

**Fig. 4-9**    *A VOM can be used to check a transistor.*

A good transistor reads like three diodes — base-emitter, base-collector, and collector-emitter — as shown. A good junction reads a high resistance in one direction and a low resistance in the other. The actual resistances will vary widely, depending on the battery voltage used in your ohmmeter, the type of transistor (silicon, germanium, etc.), and its rating. Look for the ratio between the two readings for each diode; this ratio should be high. Typical values might be 200 $\Omega$ in the forward direction and 50,000 to 75,000 $\Omega$ in the reverse.

Use the low-resistance scale on your ohmmeter, particularly in a VOM. This scale usually has a low-voltage battery (1.5 to 3 V) and won't damage the transistor. The high-resistance (megohms) scale can use up to 22.5 V, which can damage low-voltage transistors.

To check each diode, take one reading and reverse the leads to read the resistance in the other direction. If the reading is the same in both directions (100 $\Omega$ or less), that junction is shorted. For a valid test, the transistor must be disconnected; there are often very low resistances, such as coils, low-value resistors, etc., connected across the junctions. If you get a reading of infinite resistance in both directions, the junction is open. This defect is fairly rare. If you read something like 100 $\Omega$ one way and only about 500 $\Omega$ the other way, the transistor is probably leaking across that junction and should be replaced; a 5 : 1 ratio is too low.

# Voltage and resistance tests

Voltage measurements in radio and TV work are helpful. They also can be misleading—that is, unless you know what you are reading, where you're reading it, what you're reading it with, and what the reading should be! Again, all tests are comparisons against a standard, and in the case of voltage measurements the standard is the operating voltages given on the schematic diagram.

In low-impedance circuits such as the power supply, you can use any kind of voltmeter. The circuit can supply so much current that the type of meter used makes no difference. However, when you get farther along in a piece of equipment into some of the high-impedance stages, the meter impedance plays a more significant part.

Figure 4-10 shows a good example — one of the popular tube circuits used in many older TV sets as a sync separator and agc stage. Note the values of some of these resistors. In some cases, you need to read a dc voltage through a resistance of 10 or 12 M. All of the coupling capacitors, etc., that feed signals into this circuit have deliberately been left out because you are only concerned with the dc voltage relationships in this type of circuit. However, the ac signal does make a great deal of difference, not only in the output, but also in the dc voltages themselves.

Yes, this *is* a complex circuit, but you can check it rapidly if you know what to look for and the meaning of what you see. The diagram shows a set of voltages on the tube elements. Suppose you take a set of readings with a VOM. They will all be different from the values called for on the schematic. Why? Because you didn't read the instructions. Look in the corner of the schematic; it says very plainly, "dc voltages read with a *VTVM*."

For instance, if you put a VOM on a 0 V to 100 V scale (20,000 $\Omega$ per volt) and check pin 9, you won't read 100 V as the schematic calls for. The meter resistance on this scale is 20,000 $\times$ 100 =

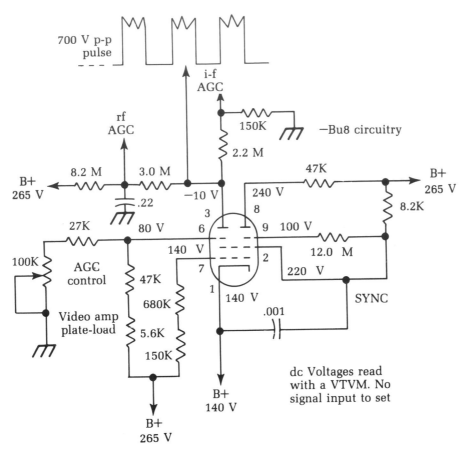

**Fig. 4-10** *One of the popular circuits used in a number of older TV sets as a sync separator and agc stage.*

2 M. Look at the size of the dropping resistor — 12 M! If you shunt only 2 M from this point to ground, you load down the circuit so that the voltage drop across the 12 M resistor is far above normal. The result is a "wrong" voltage.

If you use a VTVM, with an 11 M input resistance, you get a reading that is closer to the actual voltage. In fact, you can read the exact voltage shown on the schematic because this is the type of meter originally used to read the voltage. So, all dc voltages should match those shown if you are using the type of meter specified on the schematic. Always check the corner of the diagram. Some have voltages that were read with a 20,000 ohms-per-volt VOM. In that case, all VTVM voltages should read higher than normal.

Some circuits call for voltages that are far different from what you might expect at first glance. Notice, for example, that there is a total of 11.2 M in series with the plate of the agc section of the tube that leads to the 265 V+. According to the schematic, however, the plate should read −10 V. Why? Because a positive-going keying pulse is fed to this plate at 700 V p-p from the flyback.

There are also resistors (2.2 M and 150K) leading to ground in the IF agc circuit. The high pulse makes the tube conduct, and the plate-current flow through the load resistors makes that point negative when measured to ground. The positive voltage coming from the 265 V line through the very large resistors bucks out or balances some of the negative voltage, and you wind up with the correct agc voltage. (Part of this circuit also serves as a delay for the RF agc.)

There's one other point that can fool unwary technicians. The instructions state clearly that these voltages were read with no signal applied to the set. Why? Well, if you check up on this circuit, you'll find that there is normally a video signal voltage applied to the control grid. This signal controls the conduction *duration* of the tube—the time when plate current is flowing. The tube normally is cut off by a high bias voltage, so it isn't drawing plate current at all. When a signal fed to the grid reaches a certain value, current starts to flow, changing the voltages present considerably. Look at the monstrous resistors used; it takes only a few microamps of current to develop a large voltage across them.

Because there is no way of knowing what the actual signal input to the set would be, take the only possible standard: no signal input at all. You're now taking voltage readings under the same conditions they were taken at the factory.

## Voltage and resistance circuit analysis

In actual field servicing, operating voltages provide your best clue to trouble by changing from the normal (becoming either too low or too high). Whenever you find a tube or transistor with an incorrect voltage on any of its elements, check the complete circuit to find out why. After the voltage analysis, turn the set off and make a resistance check of all parts in the path in question, all the way back to the supply point.

For a typical example, refer back to Fig. 4-10. Suppose that either the 8.2 M or the 3 M resistor has increased in value. This occurrence would reduce the positive voltage on the plate of the tube, making the plate go too far negative. This in turn would make the agc too negative, and the controlled stages would be cut off—an agc whiteout. If the 0.22 $\mu$F bypass capacitor leaked, a low-resistance shunt path at the junction of the two large resistors would be added. Once again, there would be too little positive voltage to balance the circuit, and the same symptoms and a whiteout would occur.

How do you pin down the faulty part in such circuits? Measure the values of all resistors in the path where the fault might be. If there is too much negative voltage at the agc plate and too much positive voltage at the junction of the 8.2 M and 3 M resistors, but the 265 V supply checks normal, then suspect the 3 M resistor of being open or greatly increased in value. If there was too little (or no) voltage at this junction, suspect the 8.2 M resistor of being open.

To get a correct reading of these resistors, open the circuit. Lift one end of a resistor before making any resistance measurements across it. To read the exact value of the 2.2 M filter resistor from the plate to the IF agc, lift one end of the 150K shunt resistor to ground. If you do not, there would be at least two paths: one through the resistor and the 150K resistor to ground; the other, back through the big resistors to B+, which usually measures about 20,000 $\Omega$ to ground because of the leakage resistance of the electrolytic filter capacitors.

## Transistor circuits

In transistor circuits, you have the same basic problem with shunt paths, but with very low resistances. Figure 4-11 shows a typical class-A amplifier stage with an npn transistor. You might find such a stage in the preamplifier circuits of a hi-fi audio amplifier or in some similar type of equipment.

The signal here is developed across the load resistor, $R_L$. This resistor is of such a size that half the supply voltage will be dropped across it, and the plate current can swing above and below the operating point an equal amount; this is what makes it a class-A stage. With an 18 V supply, there's a 9 V collector potential measured to ground. (If this were a pnp transistor, it would be the same thing, but the collector voltage would be negative.)

Notice the voltage divider network, resistors R1 and R2. They are proportioned so the base will have the right bias voltage for the type of transistor used—here, 0.69 V. But the schematic says very plainly: 1.69 V! The emitter is 1 V above ground, so the actual base-emitter voltage is 1.69 − 1.0 = 0.69 V.

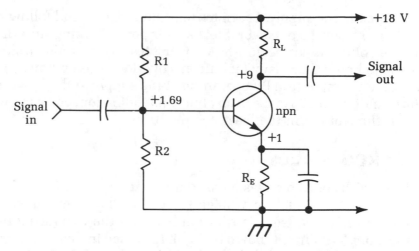

**Fig. 4-11**  *A typical transistorized class-A amplifier stage.*

In the tube circuit of Fig. 2-1, recall that both the grid and cathode of the tube read 140 V from ground. What is the actual grid-cathode voltage? That's right: $140 - 140 = 0$! In case you were wondering about the statement that the tube is normally cut off, look at pin 6. This pin is a second control grid in this type of tube and actually has a $-60$ V bias: $80 - 140 = -60$.

In such circuits, there are shunt paths again. You can't read the resistance of either R1 or R2 correctly unless you open the circuit. The resistance of the base-emitter or base-collector junction is in parallel with these resistors. Take out the transistor or lift one end of the resistor to get a correct resistance reading.

If the voltages are off-value in such circuits, don't try to pin the trouble down by doing in-circuit tests with an ohmmeter. Break the circuit up into small sections and check each part by itself. Wrong bias on the base of the transistor could be caused by a shorted transistor or by incorrect values on R1, R2, or even $R_E$.

The voltage-resistance test sequence is probably the most useful in any kind of electronics work. Properly applied, it can give you a great deal of information in the least possible time. Using improper methods can lead you farther away from the real trouble. The nonexistent defects aren't faults at all, but simply incorrect indications caused by the wrong test equipment!

In general, find the defective stage by signal tests, then check its operating voltages. If any is off value, turn the set off and find the cause of the incorrect voltage by tracing the supply circuit back to its source. Measure all resistors in this path, including the resistance of coils, chokes, and anything else that is a part of the dc path back to the supply. Learn to trace out a circuit and follow it back to the source of power. Start at this point, making sure the source voltage is correct for the very first test. You would be surprised to know how many technicians overlook this obvious and extremely simple test! By beginning at the supply and following the circuit until you come to an abnormal indication, you can pin down the trouble in the least possible time.

## Stacked stages

Circuit with stacked stages have confused many technicians because of the unusual dc voltage relations. The term *stacking* refers to the dc voltage distribution between the two (or more) transistors. The circuit shown in Fig. 4-12 is used in a commercial stereo amplifier and is a section of only one channel. It is a direct-coupled driver and output-transformerless (OTL) output stage. The circuit works in class B, with one transistor handling the positive half-cycles and the other handling the negative. The speaker is connected to their midpoint and ground.

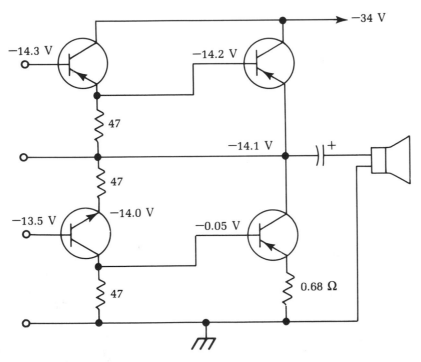

**Fig. 4-12**   *This stacked transistor circuit is from a commercial stereo amplifier.*

Notice that the maximum power supply voltage is connected to the collectors of the upper transistors and that the emitter of the upper output transistor does not go to ground, but to the collector of the lower transistor. The emitter of the lower transistor is connected to ground, which is the positive point of the power supply. (Remember, in any transistor circuit, the polarity of the power supply is determined by the type of transistor used — npn or pnp.) You can have a supply voltage of −34 V for one type or +34 V for the other and it'll work out the same. The main interest now is in the proportion of the supply voltage taken by each transistor because this proportion tells if the circuit is working properly.

Notice that the voltage does not split in the middle, as it usually does when similar vacuum-tube stages are stacked. Maximum supply voltage is −34 V, but the midpoint of the circuit has only −14.1 V. The midpoint value must be checked on the schematic so you know what is correct.

Incidentally, transistor circuits do not have the rather wide voltage tolerance that most tube circuits have. If the schematic specifies 14.1 V, it doesn't mean 14.3 or 13.8 — it means 14.1 V. In most transistor circuits, a change of only 0.2 V can cut a transistor completely off or drive it to full conduction and overheat the junction.

Learn to measure these very small voltages with extreme care. Measuring the collector current is a very good way to check the bias voltage. There is a much greater meter deflection, and it is usually easier to read on shop-type equipment.

Troubles in circuits like these are the same as in other stacked circuits: resistors that have changed in value, transistors with leakage between elements, inter-element shorts, etc. The emitter-base voltage (the bias) is a good place to start measuring if you suspect trouble. Use a sensitive and accurate VOM. A digital meter would be helpful here. Read the voltage directly from emitter to base. If this voltage is in tolerance, the transistor is probably okay. If it is off by more than 0.2 V, then there is very likely something wrong in that stage.

Note the direct coupling used in the driver transistors (the left-hand pair in Fig. 4-12). This is a very common circuit and has the same type of voltage distribution as the push-pull circuit. The available supply voltage will divide between the two in a ratio determined by the type of transistor and the bias on each one. Consult the schematic diagram of the circuit you are working on to verify the correct voltage.

# Bad base bias

In many transistor amplifier stages, a specific bias voltage will be applied to the base, regardless of the input signal. In most servicing schematics, the desired normal bias voltage will be given. Unless otherwise noted, this bias voltage value assumes that the stage's input signal is 0. For testing purposes, it is often useful to ground the input to the amplifier stage to ensure a zero signal.

Figure 4-13 shows the basic circuitry for a typical single-transistor amplifier stage. You will find this circuit (or a variation of it) in a great many electronic devices of all types.

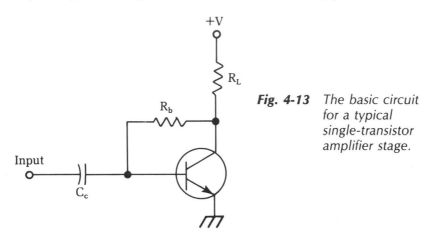

**Fig. 4-13** The basic circuit for a typical single-transistor amplifier stage.

With a zero (grounded) input, simply measure the base bias voltage by connecting your voltmeter between the base of the transistor and circuit ground. You should get the value given in the schematic or other troubleshooting literature for the equipment being serviced. If this voltage is off, the first thing to suspect is the supply voltage. (The tolerable error will depend on the intended application and the precision required for the particular circuit you are working on.)

If the supply voltage is okay, check the transistor's collector. If this value is way off the nominal value (from the schematic or servicing literature), you might have a bad transistor. Refer to chapter 8 for information on testing transistors.

Assuming the transistor's collector voltage is okay, but the base bias voltage is incorrect, there are two likely possibilities: either the coupling capacitor ($C_c$ in Fig. 4-13) is leaky or there is some unintended leakage resistance between the transistor's collector and base. This leakage resistance might be internal or external to the transistor itself.

To narrow down the cause of the problem, it is usually easiest to check out the coupling capacitor for leakage first. To do so, make sure the input to the amplifier stage itself is not grounded. The input to an earlier stage in the circuit can be grounded if necessary to force a zero input condition. Now, monitor the transistor's base voltage as you short-circuit the input end of the capacitor to ground. If the measured voltage changes, especially if by a large amount, then the coupling capacitor is almost certainly leaky and should be replaced. On the other hand, if there is no effect on the measured base voltage, the capacitor itself is probably not the source of the problem in the circuit. The defect must be collector-to-base leakage of some sort.

Unfortunately, it takes a little more effort to trace a collector-base leakage problem because you must do some desoldering. Disconnect one end of the base resistor ($R_b$). It would probably be a good idea to check out the resistance of this component with an ohmmeter while it is disconnected. If this resistor has somehow changed value significantly, it could be the cause of the problem.

To test the transistor for internal collector-to-base leakage, monitor the collector voltage (from the transistor's collector to circuit ground). Now temporarily short together the base and emitter of the transistor. If the transistor is good, the collector voltage should jump up to the full supply voltage. (Don't be misled by any circuit resistances between the actual power supply and the transistor's collector.) If the measured collector voltage with the emitter and base shorted is less than the expected supply voltage, then the transistor is internally leaky. Replace it.

# Dealing with noise pick-up

If you are working in an area with very strong electromagnetic signals, such as radio waves, or if very high level signals are used in one part of the circuit you are working on, you might find yourself plagued with noise pick-up problems when you attempt to take measurements. The test probes and leads can act like antennas, picking up any strong nearby electromagnetic signal. Such unwanted signals also can sneak in through ac power lines.

This unintentionally picked up noise signal will be added to the signal you are attempting to measure. If the interfering signal is quite strong and the signal you are trying to measure is small, the desired signal might be completely engulfed and lost in the noise. Obviously, you can't get an accurate, or even a meaningful, measurement under such conditions.

Battery-powered test equipment tends to be somewhat less prone to such noise pick-up problems. When you encounter this type of trouble, switch over to battery power if possible. Disconnect the ac adapter. This action will reduce line induced noise, which is one of the most common sources for this type of problem.

If you must use ac power for your test equipment, or if switching over to battery power doesn't sufficiently clear up the problem, try connecting a 0.1 $\mu$F capacitor across the multimeter's input jacks. This solution usually will filter out the picked up noise signal quite well. For best results, use a polyester capacitor, or something similar.

# 5
# Oscilloscope tests

A TYPICAL SERVICE-TYPE OSCILLOSCOPE WILL NOT READ DC VOLTAGE, dc current, ac current, or even ac voltage—directly. (You can read ac voltage on a scope, but only by comparing the vertical deflection to the vertical deflection of a known ac voltage from a calibrator.) Despite this handicap, the scope is the handiest instrument in the whole shop. It can do something no other instrument can—read signals. In one of the most simple tests—using a test probe to touch one point and then another in a circuit—the scope can immediately and definitely depict if a given stage has any gain. Because a lot of time is spent trying to locate certain stages with insufficient gain, the scope can be a real timesaver. In many cases, it is the only instrument that can give the necessary information.

An ac voltmeter can determine the ac voltage in a circuit, but the scope is the only instrument that determines whether the voltage is hum, vertical sync, video, audio, or some other type of waveform. By comparison with a voltage standard (calibrator), it also depicts amplitude.

Make gain checks with ease: feed in a test signal of the frequency normally used in the circuit and check the signal levels at the input and output of any amplifier stage. If you read a vertical deflection of 1 unit on the input and 8 units on the output, the stage has a voltage gain of 8.

The scope is the closest thing to a really universal test instrument. It works with tubes, transistors, printed circuits, integrated circuits, or any other kind of circuitry with equal ease because it reads signals, the one thing all circuits have in common. No matter how a circuit is built or what voltages it uses, it handles signals — amplifies them, clips them, shapes them, or does something to them — and you can check these actions with the scope. Use the scope in radio and TV work as often as the VOM or VTVM; it really speeds up the work.

Let's examine a few typical scope patterns — learn to recognize and interpret them. Like most tests, the important question isn't "What have we here?" but "What does it *mean?*" As an example, consider the scope display shown in Fig. 5-1.

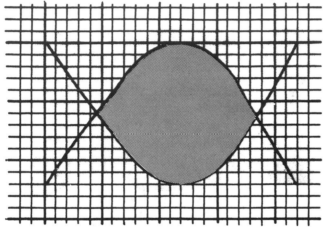

**Fig. 5-1**  *An oscilloscope with sweep set-off frequency.*

Notice that the pattern height covers 12 small squares on the calibrated screen. The vertical deflection represents an ac signal, but the scope sweep is set off-frequency, so you can't see the individual cycles of the signal. It could be a sine wave, square wave, or anything.

This scope image can be used, however. Suppose you're feeding an audio signal into the input of an amplifier and touch the probe to the circuit. If you see this pattern on the input of a stage (tube grid or transistor base), then move the probe to the output (tube plate or transistor collector) and see the same pattern (height and all), the stage has a gain of 1.

What this gain signifies depends on what you're testing. The scope reads VOLTAGE. If you are testing the preamplifier stage of a tube amplifier, this stage should have a very high voltage gain (up to 50). If, however, you are testing a transistor preamp stage, then a 1:1 voltage ratio might be perfectly normal. Such transistor stages often have a 1:1 voltage ratio, but a large current gain that gives the required gain in signal power. For a voltage ratio of 1:0.25, it's trouble.

If you are looking for distortion, you do need to see the individual cycles. Feed in the signal and touch the probe to any point in the circuit where the signal can be seen (preferably at the input) so that you know what the test signal looks like before starting. Then adjust the scope's horizontal sweep to see the individual cycles, as illustrated in Fig. 5-2. Adjust the sync-lock control of the scope to stop the pattern. This is a fairly good sine-wave pattern, but you need to get a better look at the waveshape. To do so, spread it out a little by using a faster sweep or turning up the horizontal-gain control.

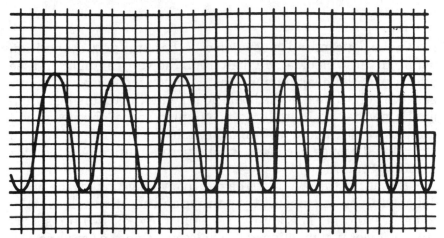

**Fig. 5-2**    *The oscilloscope's horizontal sweep control is adjusted to see the individual cycles.*

If the waveform shows distortion and you suspect there is signal clipping somewhere in the amplifier, trace the signal through the amplifier from the input until a pattern similar to the one shown in Fig. 5-3 appears. Figure 5-3 shows a decided clipping on one-half of the sine-wave signal. By checking the circuit, you can tell which part is most likely to cause this distortion. In certain circuits, this could be the correct waveform. In a transistor class-B output stage, each transistor carries one-half of the sine-wave signal. If, however, the output at the speaker looks like Fig. 5-3, only one-half of the circuit is working, so look for trouble there.

You can identify positive and negative halves of a sine wave. You only need to know which way the scope deflects on a positive-going signal. Most scopes have a reversing switch that will

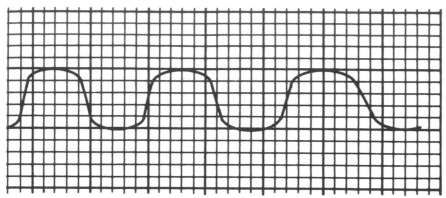

**Fig. 5-3**    *This sine wave signal is severely clipped on one half-cycle.*

invert the pattern, and this switch must be in the NORMAL position before the test. Hook a 1.5 V battery to the vertical input and see which way the beam jumps. Most service scopes are set up so a positive voltage makes the beam go up with the reversing switch in the normal position. In an ac-coupled scope, the beam will jump up or down and then come back to the original position. In dc-coupled scopes, the beam will be permanently deflected by a specific amount.

By setting the switch so the positive voltage makes the beam deflect up, you can identify the positive and negative halves of the signal waveform. This ability is useful in the class-B transistor-output stages just mentioned. You also can find the *zero line*, the position where the beam rests when there is no signal input. By adjusting this position to the center line on the calibrated screen, you can see how much of a waveform is positive and how much is negative.

It is not imperative for a scope pattern to be clear and sharply focused to be useful. In the horizontal-oscillator tests later you can use a pattern like the one shown in Fig. 5-4. What you are looking for in this case are horizontal sync pulses in a TV signal. The illustration shows three of them; the third is at the extreme right of the pattern. (Our scope sweep, therefore, is running at 15,750 Hz divided by 3, or 5,250 Hz. Use this setting to make comparison tests anywhere in the horizontal sweep circuits.) Don't waste time trying to make picture-book patterns when all you need to know is the number of cycles of signal that can be seen on the scope.

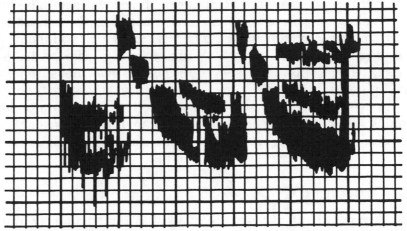

**Fig. 5-4**  *This oscilloscope display includes the horizontal sync pulses in a TV signal.*

# Calibration for direct measurements

The height of the pattern on the scope screen depends on the setting of the vertical gain control. A 2-inch pattern means nothing unless you have previously set the vertical gain for a 2-inch deflection when a known voltage (1 V p-p or 10 V p-p) is applied to the input. One way to set the gain is with an external voltage calibrator, which is just a tapped transformer with an accurate ac voltmeter connected across the output terminals. The scale on these meters is usually calibrated in rms, peak, and peak to peak voltage. A given deflection can be read as any one of three, depending on what you want it to mean in a particular test. *Unless a special probe is used, a scope always displays a peak to peak voltage.*

Many scopes have built-in voltage calibrators. Some have variable calibrating voltages with a volt-reading knob. Others have a regulated 1 V p-p output. By using the calibrated step attenuator, set the scope for 1 V on the lowest (most sensitive) range and turn the attenuator to ×10 and read 10 V for the same deflection; to ×100 and read 100 V for the same deflection; etc. Check the instruction book for your scope to see how this attenuator is marked. Some are marked in gain and others in attenuation. They do the same thing; the markings are just different.

If you don't have a scope calibrator, you can use any kind of variable ac voltage, measuring the value with an ac voltmeter (remember that the standard ac meter reads *rms* voltage). The filament voltage of your tube tester is a very useful calibrator, since it runs from 1.1 V to 117 V. It, too, is in rms values, and you'll have to use the conversion formula: 10 V rms = 14.14 V peak = 28.28 V peak to peak. If you want a very rough peak to peak measurement, just figure that the peak to peak is three times the rms voltage.

As you go through the various scope tests to speed up electronics servicing, the scope you will use mainly to answer these questions: "Is it there, or isn't it; if it is there, how big is it?"

# Oscilloscope probes

The purpose of any test is to discover what's going on in a circuit. During a test, you don't want to disturb the circuit any more than is absolutely necessary. Hence, we prefer the high-resistance VTVM to the VOM. A VOM could have as little as 40,000 Ω input resistance (on a 2 V scale, 20,000 Ω per volt).

The most sensitive instrument in the average shop is the oscilloscope. Its normal input has a very high resistance — up to several megohms — and a very small shunt capacitance — only a few picofarads. The direct input of a scope can be used for many tests as is. It also can be used in any audio-frequency circuit, most video-frequency circuits, for ripple testing, etc.

There are specialized probes to get information concerning special circuits — tuned circuits, very high resistance circuits, etc. — where a minimum of capacitance and a maximum of resistance is needed to keep from upsetting circuit conditions when test equipment is hooked up. These special probes have cylindrical housings with test tips of various kinds and are attached to the vertical input of the scope through well-shielded cables to keep them from picking up stray signals — for example, radiation from the horizontal-sweep circuit, hum, etc. The probe bodies usually are shielded as close to the tip as possible for the same reason, and the ground lead is usually kept very short and attached directly to the probe body.

## Low-capacitance probe

The low-capacitance probe is perhaps the most popular of the specialized probes. Figure 5-5 shows the schematic of a typical unit. Notice that there are actually two capacitors in series in the input. The 0.05 $\mu$F serves mostly as a dc blocking capacitor. Low-capacitance probes are used at high frequencies, and the series reac-

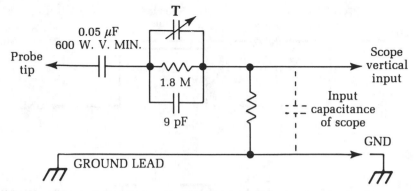

**Fig. 5-5**   *This low-capacitance probe is probably the most popular type of specialized oscilloscope probe.*

tance of a 0.05 $\mu$F capacitor is so low at such frequencies that you usually can neglect it. The capacitor that counts is the little 9 pF unit shunted across the 1.8 M resistor.

This capacitor, in series, actually forms an ac voltage divider with the input capacitance of the scope. The size of the capacitor and the value of the cable capacitance are so proportioned that the probe has a 10:1 step-down ratio; it is called a *divider probe*. Therefore, if you see a 1 V peak to peak deflection on the scope, the actual signal voltage is 10 V p-p.

The little trimmer capacitor shown above the resistor is used to make fine-tuning adjustments of the total probe series capacitance in order to match it exactly to the input capacitance of the scope itself and to the cable capacitance. Proper matching is needed to obtain the right division ratio and to sharpen the high-frequency response. The resistors used here are all big ones, and their exact values are not usually crucial.

### Detector probe

The detector probe seems to be next in popularity. A scope won't show the modulation of a radio-frequency signal until it has gone through the set's detector circuit — the video detector in TV sets, the second detector in radio receivers, etc. If you need to read the signal level, modulation, etc., of a TV or radio signal in the IF stages, tuner, and other circuits ahead of the detector, you must provide your own detector. Figure 5-6 shows schematics of two detector probes. Of them, the shunt type seems to be used most often, but either one will work. All they do is demodulate the signal so that you can see what's happening to it in the stage under test.

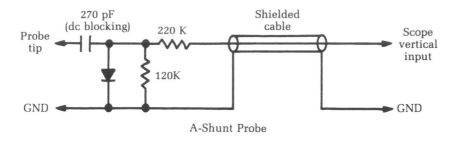

A-Shunt Probe

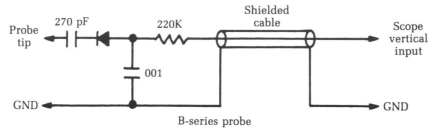

B-series probe

**Fig. 5-6** *Two detector probe circuits.*

Each probe consists of a crystal-detector diode, with a couple of resistors and capacitors. For safety's sake, a dc blocking capacitor usually is placed between the diode and the probe tip in case you should accidentally touch a high-voltage point in the circuit. A voltage of 200 to 300 V dc wouldn't do a 1N64 diode much good.

You can use a detector probe for signal tracing through the video IF of a TV set if you feed in an AM signal. The special tests you can make with this type of probe are explained as we go along. There are also voltage-doubling probes with two diodes, but most detector probes are simple half-wave rectifiers. The type of circuit used is usually marked on the probe body.

## Resistive-isolation probe

There's one more specialized probe: a resistive-isolation type. It is a resistor, somewhere around 50,000 Ω, mounted in a probe body. This probe is used in some tuner tests, video-amplifier tests, etc., so you get a little more resistance and can isolate the scope's input capacitance from the circuit as much as possible.

## Direct probe

The direct probe is just what it says: the wire from the vertical input of the scope comes straight through the probe body to the tip. The body of the probe is, or certainly should be, well shielded as close to the tip as possible to hold the pickup of stray signals to a minimum.

# Triggered and gated sweep

The horizontal (X-axis) and vertical (Y-axis) signals in an oscilloscope must be synchronized, or the display will be a hopelessly indecipherable muddle. Most early oscilloscopes, and some low-cost modern oscilloscopes, use a synchronization method called *recurrent sweep*. A free-running oscillator built into the oscilloscope's circuitry generates the sawtooth-wave (or ramp wave) sweep signal. The Q2 of this sweep signal is manually adjusted until a stable display is achieved, indicating that the sweep has been properly synchronized—certainly the most direct and obvious way to accomplish the task.

The recurrent-sweep method of synchronization is functional, and it has the advantage of being easy and inexpensive to implement. It is, however, extremely limited for much practical use. It is very difficult, and in some cases impossible, to change the beginning trigger point of the display to view a specific portion of a long waveform. It is also very difficult to determine precisely the actual period of the sweep signal when recurrent sweep is used because the sweep frequency is actually altered to achieve synchronization. This alteration will inevitably throw off any frequency or timing measurements you attempt to make with the oscilloscope.

Recurrent synchronization also won't work for displaying transient or aperiodic signals, such as an occasionally activated switching voltage. Long before you can adjust the sweep frequency control for proper synchronization of the display, the signal of interest will be long gone.

For these reasons, among others, the simple and "obvious" approach of sweep synchronization is rarely used in modern oscilloscopes beyond a few of the simplest bottom-of-the-line models intended primarily for beginning hobbyists and experimenters.

A much better approach to oscilloscope synchronization is known as *triggered sweep*. Instead of a free-running sawtooth-wave oscillator, a triggered-sweep oscilloscope uses a one-shot (timer) circuit with a ramp output. One complete ramp cycle is generated each time the one-shot circuit is triggered. If there is no trigger signal, no ramp cycle will be generated.

The time period of this ramp can be set with a single control. On most, but not all, oscilloscopes, this control is labeled VOLTS/DIV (volts per division), or something very similar.

Notice that the actual time period of the ramp is not affected by the trigger signal in any way. The trigger is just used to determine when the ramp's timing period will begin. If, for example, the VOLTS/DIV control is set to give a ramp period of 1 second, but the trigger pulses come every 1.5 second, the result will be a 1-second ramp, followed by a 0.5-second pause before the next 1-second ramp begins. This is a very extreme example for purposes of illustration.

The triggering can be synchronized to any point in the input wave cycle or to a discrete, one-shot event of some sort. Thus, the display can begin at any desired instant. This type of synchronization usually is offered in two selectable modes: *automatic synchronization* (auto sync), and *normal synchronization*.

Somewhat curiously, considering the names used for these two synchronization modes, the automatic synchronization mode is more commonly employed in practical electronics work than the so-called normal synchronization mode.

In automatic synchronization, the sweep signal is triggered each time the input signal (being monitored) passes through a baseline reference point. This point usually will be 0 V, but it can be adjusted to account for any dc voltage offsets. In this mode, a trace is displayed at all times, whether an input signal is present or not. This feature makes this mode particularly useful for monitoring weak signals with an a amplitude that is too low for reliable direct triggering.

In the normal synchronization mode, the sweep ramp is triggered after the input signal passes through a specific preset voltage. There is no imaginary baseline reference, as in the automatic synchronization mode. Therefore, it is important that the vertical (input) waveform crosses this preset level, or the sweep circuit will not trigger.

On most oscilloscopes, a level control can be used to shift the voltage of the triggering point above or below the nominal 0 V level, permitting even negatively biased waveforms to properly trigger the sweep circuit. In a dual-trace oscilloscope (refer to chapter 1), the trigger can be taken from either channel A or channel B, or it can alternate between the two.

Another type of synchronization used in some oscilloscopes is called *line-sync triggering*, in which the sweep waveform is triggered by the 60 Hz power line. Obviously, this type of synchronization is only suitable for fairly low-frequency signals that are time-related to the standard ac power line frequency.

Many modern oscilloscopes also permit external triggering. With this option, an external trigger signal (which can be almost anything) is used to drive the oscilloscope's one-shot circuit to generate the sweep ramps at the appropriate times. The frequency of this trigger signal should be related harmonically to (a whole number multiple of) the signal to be displayed, or the oscillator will be out of sync and the oscilloscope's screen will just display a lot of meaningless garbage.

# Measuring dc voltages with an oscilloscope

The oscilloscope is designed primarily for the measurement and analysis of ac voltages, but it can be used to measure dc voltages, as well. Generally speaking, it is usually more convenient and more accurate to use a multimeter to take dc readings, but occasionally when you are working with an oscilloscope, you'll find that you need to make a quick check of a dc voltage or two. In this case, it might be less trouble to just use the oscilloscope, rather than switch over to a different test instrument in midstream.

Throughout the remainder of this chapter, assume that all tests are performed on a single-trace oscilloscope, unless otherwise noted. All dual-trace oscilloscopes can be used as if they were single-trace scopes. Many, if not most, of the single-trace tests can be performed just as well in the dual-trace mode. The only real difference is the position of the zero baseline on the display screen. For a single-trace oscilloscope there is just one baseline, and it is usually positioned halfway down the CRT screen.

To measure a dc voltage with an oscilloscope, you must use the automatic synchronization mode. You'll get no display at all in the normal mode. An oscilloscope with recurring sweep or line-sync triggering also will work for dc voltage measurements.

Before making a voltage measurement with an oscilloscope, first check to make sure the zero baseline is exactly where you think it is. Manual controls are provided to move this baseline in either the positive or negative direction. If you think the zero baseline is at the center of the screen, but it has actually been offset downward 1.5 squares, you obviously won't get accurate voltage readings.

First, calibrate the offset control with no signal applied to the

oscilloscope's test probe. For the cleanest, most noise-free adjustment, it is usually helpful to ground the test probe while calibrating the offset. You should see a straight, featureless line across the
width of the CRT screen. This is the zero baseline. Adjust the
oscilloscope's offset control until the displayed baseline is right on
the centermost grid line of the CRT screen. You are now ready to
make the dc voltage measurement.

Connect the ground clip of the test probe to the circuit ground
or common point (0 V reference) of the circuit under test. Then
touch the probe tip to the desired test point within the circuit. If a
voltage is present, the displayed line will jump up or down from
the center line. Assuming the signal is a pure dc voltage, it will
still be a straight, featureless horizontal line. It has simply been
displaced in the vertical dimension by a specific amount proportionate to the voltage value. If the displayed line has moved upward from the baseline position, then you have a positive dc voltage; if the displayed line has moved downward from the baseline
position, then you have a negative dc voltage.

To determine the magnitude of the signal voltage, simply
count the number of division lines between the displayed signal
line and the original center line (baseline). If the displayed line is
not exactly on a grid line (and it rarely will be), estimate how
much of a division is included—one-half, one-third, or whatever. Then look at the setting of the volts-per-division control on
the oscilloscope's front panel. Multiply the number of grid divisions you counted by the volts-per-division factor. That is:

$$V_{in} = DIV \times \frac{V}{D}$$

For example, let's say the displayed line is 2.5 divisions above
the baseline, and the VOLTS/DIV control is set for 0.5 V/division.
In this case, the monitored voltage has a value of approximately:

$$V_{in} = 2.5 \times 0.5$$
$$= 1.25 \text{ V}$$

Here's another example. Let's say the displayed line is 3.75
divisions below the baseline. Since it is below the original baseline, you know you are dealing with a negative voltage. This time
the VOLTS/DIV control is set for 10 V/division, so the monitored
voltage this time is equal to:

$$V_{in} = -3.75 \times 10$$
$$= -37.5 \text{ V}$$

The accuracy of such dc voltage measurements obviously is limited by how well you can determine the fractions of divisions included in the measurement. For maximum accuracy, use the smallest V/division setting that will leave the signal on the readable portion of the CRT screen. If the signal voltage is too large for the V/division setting, the displayed line will be "drawn" above or below the edges of the screen and you won't be able to see it. No harm will be done to the oscilloscope itself under such conditions, unless the voltage is very large, exceeding the instrument's maximum specifications.

# Measuring ac voltages with an oscilloscope

The oscilloscope was designed primarily to work with ac waveforms, but the procedure for measuring an ac voltage is basically very similar to that for measuring a dc voltage. With ac voltages, however, the oscilloscope's graphic display offers a lot of additional information. For the time being, we will just concentrate on the signal's amplitude, or the ac voltage.

When you are measuring ac voltages and are calibrating the offset control, make sure the oscilloscope is set for the automatic synchronization mode. Once the offset control has been properly adjusted so that the no-signal baseline is displayed on the centermost grid of the oscilloscope's screen, you can switch the instrument's mode control to the normal synchronization mode if that is suitable for the particular tests you intend to run.

Connect the ground clip of the test probe to the circuit common or ground (0 volt) point of the circuit under test. Then touch the probe tip to the desired connection point within the circuit. If an ac voltage is present, a waveform will be displayed on the CRT screen. At this point, you might see one or two cycles of the waveform (the sweep frequency is correctly set); you might see just part of a waveform cycle (the sweep frequency is too low); or you might see a multitude of cycles crammed too close together to be decipherable (the sweep frequency is set too high). Since all we are interested in is the amplitude of the signal, the details of the waveform aren't too important. You should, however, make sure you have at least one complete cycle displayed on the CRT screen. Adjust the TIME/DIV control if necessary. At this stage, you don't have to worry about setting this control precisely.

Usually, the displayed waveform is centered around the no-signal baseline. In some cases, however, you might find the baseline of the ac waveform is shifted up or down the screen. If this happens, the ac signal you are measuring is riding on a dc offset voltage. First measure the dc voltage as described in the preceding section. (Ignore the ac waveform. All you are interested in is the central baseline of the ac signal and the amount it has shifted from the original calibration (no-signal) position.) Once this is done, you might want to readjust the oscilloscope's offset control to center the ac waveform on the screen.

To determine the magnitude of a test voltage, count the number of division lines covered by the displayed waveform. If this waveform does not cover an exact whole number of divisions (which is usually the case), estimate how much of a practical division is included within the range of the waveform. Then, look at the setting of the V/division control of the oscilloscope. Multiply the number of divisions you counted by the volts-per-division factor.

Of course, with an ac waveform, the instantaneous voltage is changing continuously over time. At what point in the cycle do you measure the voltage?

One obvious approach is to measure the peak value of the ac waveform — when the signal is at its maximum distance from the baseliner. This value is often easiest to see when the sweep frequency is set so a great many cycles are displayed at once and the details of the waveform are obscured, as illustrated in Fig. 5-7. However, you can also easily find the peak level when just one or two cycles are displayed with clear detail, as shown in Fig. 5-8. The peak voltage is measured from the center baseline, which represents the 0 V reference point.

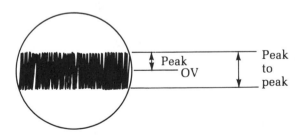

**Fig. 5-7**   *The peak voltage of an ac signal is often easiest to see when many cycles are displayed with the details obscured.*

Although a peak measurement is quite useful for certain purposes, such as determining the absolute maximum ratings of a component or a circuit, this type of measurement is somewhat misleading. As an example, let's say the sine wave in Fig. 5-8

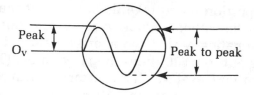

**Fig. 5-8**  *When just one or two cycles are displayed, the peak voltage is the maximum deflection from the zero line.*

reaches a maximum (peak value) of 10 V. Notice that the signal line is never farther than 10 V from the baseline, although for much of the cycle, there is less displacement than this maximum.

Because the sine wave is symmetrical around 0 (the baseline), it also has a negative peak of −10 V. This is the maximum deflection below the baseline. Some complex waveforms might have unequal positive and negative peak voltages.

Obviously, to say we have 10 V ac is rather misleading because the signal voltage actually reaches the full 10 V for only a brief instant during each cycle. The rest of the time the actual instantaneous signal voltage is less than 10 V. This ac signal will be considerably weaker and won't be able to do nearly as much work as a 10 V dc signal. Also, common electronics calculations like Ohm's law won't work with peak values.

Because the signal voltage varies between a +10 V positive peak and a −10 V negative peak, there is a 20 V difference between the two peaks. That is, the signal voltage varies over a 20 V range. Therefore, we can say that this magnitude of the signal is 20 V *peak to peak*. Again peak-to-peak measurements are sometimes useful for certain purposes, but it would be very misleading to say that we have 20 Vac. Clearly, this signal wouldn't come close to being comparable to a 20 Vdc signal.

It might seem logical that you could take an average of the instantaneous values passed through the complete cycle and come up with a meaningful average voltage. Unfortunately, with a symmetrical waveform like the sine wave, you wouldn't get a useful result at all. The positive portion of the cycle is a mirror image of the negative portion. For every positive instantaneous voltage, there will be an equal negative instantaneous voltage. All the individual values will cancel each other out, leaving you with an average value of 0, regardless of the amplitude of the ac signal. Clearly, this is the most useless type of voltage measurement.

A reasonable compromise for symmetrical waveforms is to use only half the complete cycle — either the positive portion or the negative portion—and take the average of that. This measurement can be rather tedious to work out, but it has been proven mathematically that the average value of a sine wave always works out to 0.636 times the peak value. (This does not work for any other waveform. It is true for sine waves only.) So, in our example, if the peak voltage is 10 V, then the average voltage is:

$$Vac_a = V_p \times 0.636$$
$$= 10 \times 0.636$$
$$= 6.36 \ Vac_a$$

where $Vac_a$ is the average ac voltage, and $V_p$ is the peak voltage.

Conversely, if you know the average value of an ac signal (sine waves only) and need to find the peak voltage, you can multiply the average voltage times 1.572327. For most practical electronics work, this can usually be rounded off to 1.57, or even 1.6.

Although the average voltage does give you a fair idea of how much voltage is being passed through the circuit, you still must remember that common electronics equations like Ohm's law don't hold true with this type of value. As a result, many circuit design calculations are difficult, or even impossible, to perform. What you really need is a way to express the ac voltage in terms that can be compared directly to equivalent dc voltage. In other words, you want 10 Vac to cause the same amount of heat dissipation in a given resistor as 10 Vdc.

You can find such an equivalent value by taking the rms of the sine wave. The mathematics involved are rather complex, but as long as you are working with sine waves only, it has been demonstrated that the rms value works out to 0.707 times the peak value. This is not true for other waveshapes.

For example, we have an ac voltage of:

$$Vac_{rms} = V_p \times 0.707$$
$$= 10 \times 0.707$$
$$= 7.07 \ Vac \ (rms)$$

The rms value is the one most commonly used in the majority of practical ac measurements. By use rms values, Ohm's law and other important calculations can be used in exactly the same way as in dc circuits.

It is very important to realize that these equations hold true only if the ac waveform is a sine wave. You won't get correct results with other waveshapes. For comparison purposes, however, most non-sine-wave ac signals can be measured as if they were sine waves. You won't get strictly accurate results, but they usually will be close enough for most practical electronics work.

For example, a schematic might indicate that a square wave signal at a certain point in the circuit should measure 12 Vac rms. The original reference measurement was probably made with a standard ac voltmeter, calibrated for sine waves. If you measure the signal with a similar meter or use an oscilloscope and take 0.707 of the measured peak voltage, you should get a value of about 12 V, too. This compromise sidesteps the need to recalibrate all ac test equipment every time the waveshape changes, or to perform complicated mathematical calculations for every measurement.

You can even get a rough idea of just how far off the measured values will be when you use an oscilloscope. The closer the waveshape in question is to a sine wave, the more accurate the measurement will be.

For your convenience, the various ways to measure the amplitude of an ac (sine wave) signal are compared in Table 5-1. These same relationships hold true for ac currents, as well as ac voltages.

**Table 5-1   The amplitude
of an ac (sine wave) signal
can be measured in various ways.**

| | |
|---|---|
| RMS | $= 0.707 \times$ peak |
| RMS | $= 1.11 \times$ Average |
| Average | $= 0.9 \times$ RMS |
| Average | $= 0.636 \times$ peak |
| Peak | $= 1.41 \times$ RMS |
| Peak | $= 1.57 \times$ Average |
| Peak | $= 0.5 \times$ peak-to-peak |
| Peak to peak | $= 2 \times$ Peak |
| Peak to peak | $= 2.82 \times$ RMS |
| Peak to peak | $= 3.14 \times$ Average |

# Measuring frequency
# with an oscilloscope

The oscilloscope is not just a glorified voltmeter. It also can be used to measure the frequency of any periodic ac waveform.

The first step in measuring frequency with an oscilloscope is to adjust the sweep frequency and the scope's trigger controls to get a clear, stable display of the monitored waveshape. The sweep control on most modern oscilloscopes is calibrated with time-per-division markings. For example, a time base of 2 mS (0.002 second) per division can be selected, in which case it will take the oscilloscope 2 mS to draw the trace from one vertical grid division line to the next.

The next step is to count the number of vertical divisions on the oscilloscope's screen from the beginning of one complete cycle to the beginning of the next. It is helpful if you can adjust the sweep frequency so one complete cycle of the displayed signal covers a whole number of divisions, but at times you might need to estimate the partial division left over at one end or the other — one-half, one-third, or whatever.

As an example, let's say one cycle of the displayed waveform covers about 4.3 divisions. Multiply this number by the timebase time-per-division value to get the overall time period of the measured signal's cycle:

$$T = D \times t_d$$

where $D$ is the number of vertical divisions, and $t_d$ is the time-per-division setting of the oscilloscope's sweep control. In our example, this works out to approximately:

$$T = 4.3 \times 0.002$$
$$= 0.0086 \text{ second}$$
$$= 8.2 \text{ mS}$$

Finally, to find the actual signal frequency, take the reciprocal of the time period of one complete cycle. That is, divide 1 by the time period:

$$F = \frac{1}{T}$$
$$= \frac{1}{D \times t_d}$$

The time period must be in seconds for this equation.

For our example, we find that the signal frequency works out to about:

$$F = \frac{1}{0.0086}$$
$$= 116.3 \text{ Hz}$$

These frequency equations can be used for any periodic (repeating cycle) ac waveform. Since the pattern is continuously repeating, you can select any convenient point as the "beginning" of a cycle, as long as you are consistent from cycle to cycle. Usually, it is most practical to assume the nominal beginning point of the cycle is the point where the signal crosses the baseline (zero), moving in the positive direction. Be careful, though. Some complex waveforms might cross back and forth through the baseline more than once per cycle. An example is shown in Fig. 5-9.

You can use the same basic procedure described here to determine the timing of an aperiodic (one-shot) electrical event, such as a change in a switching voltage.

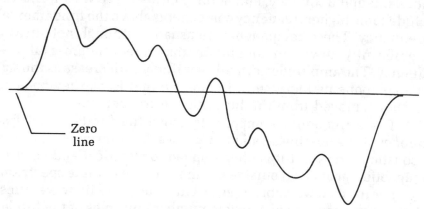

**Fig. 5-9**  *Some complex waveforms might cross back and forth through the baseline more than once per cycle.*

# Analyzing distortion with an oscilloscope

Because an oscilloscope permits the electronics technician to directly view the waveshape of the signal being monitored, any distortion of a known waveform can be spotted easily. Distortion, by definition, involves any undesired change in a waveform. If the waveform displayed on an oscilloscope doesn't look the way it is supposed to look, the monitored signal is presumably being distorted somewhere along the line. *Note:* often improper test procedures can make the displayed waveform look distorted, even though the signal actually passing through the circuit under test is fine.

Distortion is often the primary cause of malfunctions in electronic equipment. If the circuitry cannot recognize a control signal because of excessive distortion, it won't be able to respond correctly. Of course, distortion is highly undesirable in audio applications, such as in amplifier circuits.

## Harmonics

To get a better idea of distortion and its effects, you will need to understand the concept of harmonics. Most waveforms, with the exception of a pure sine wave, are composed of multiple-frequency components.

The nominal base frequency of any waveform is called the *fundamental frequency*. This is the basic cycle repetition rate of the waveform. Except for the sine wave — which consists of the fundamental frequency, and nothing else — all ac waveforms include some higher frequency components above the fundamental frequency. These components are usually, although not always, significantly lower in amplitude than the fundamental frequency. The amplitude normally continues to decrease as the signal components increase in frequency; that is, as they become farther removed from the fundamental frequency.

These frequency components above the fundamental frequency are sometimes called *overtones*. This term comes from acoustic theory, but it is often applied to electrical and electromagnetic signals well outside the limits of the audible spectrum.

For most ac waveforms, especially true periodic waveforms, these overtones are exact whole number multiples of the fundamental frequency. This type of overtone is called a *harmonic*. Some waveforms do include overtones that are not exact whole number multiples of the fundamental frequency. These irregular overtones are called *enharmonics*. Any waveform that contains enharmonics will inevitably change its waveshape at least a little from cycle to cycle.

Common electronic waveforms used for testing purposes are almost always truly periodic and contain no enharmonic content. Therefore, this discussion will concentrate solely on harmonics.

The fundamental frequency component of a complex waveform can be thought of as a simple sine wave. Each successive harmonic is like an additional sine wave at a higher frequency that is an exact whole number multiple of the fundamental frequency.

As an example, let's consider a signal with a fundamental frequency of 350 Hz. Possible harmonics that might be included in this waveform are as follows:

| 350 Hz | fundamental | |
|---|---|---|
| 700 Hz | 2nd harmonic | 2 × fundamental |
| 1050 Hz | 3rd harmonic | 3 × fundamental |
| 1400 Hz | 4th harmonic | 4 × fundamental |
| 1750 Hz | 5th harmonic | 5 × fundamental |
| 2100 Hz | 6th harmonic | 6 × fundamental |
| 2450 Hz | 7th harmonic | 7 × fundamental |
| 2800 Hz | 8th harmonic | 8 × fundamental |
| 3150 Hz | 9th harmonic | 9 × fundamental |
| 3500 Hz | 10th harmonic | 10 × fundamental |

and so forth. This series can be continued indefinitely. Generally speaking, the higher the harmonic, the weaker its amplitude, so you often can simplify matters by ignoring any harmonic content above some reasonable cut-off point. Any harmonics above this point are considered to have too weak an amplitude to make a substantial impact on the noticeable characteristics of the waveform.

Not all waveforms contain all the possible harmonics. As already mentioned, a sine wave (shown in Fig. 5-10) consists of just the fundamental frequency. It has no harmonic content at all. Any harmonic content in a sine wave is a form of distortion.

**Fig. 5-10** *A sine wave consists of just the fundamental frequency, with no harmonic content.*

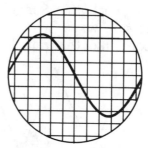

A sawtooth wave, like the one illustrated in Fig. 5-11, contains all the available harmonics at relatively high amplitudes, in addition to the fundamental frequency.

A square wave, like the one shown in Fig. 5-12 consists of the fundamental frequency and only the odd harmonics. All the even harmonics (second, fourth, sixth, etc.) are omitted from this waveform. For example, a 350 Hz square wave would have the following harmonic make-up:

| Hz | Harmonic |
|---|---|
| 350 | fundamental |
| 1050 | 3rd harmonic |

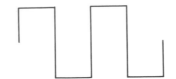

**Fig. 5-11**   *A sawtooth wave contains all the harmonics at relatively high amplitudes.*

**Fig. 5-12**   *A square wave has all the odd harmonics, but no even harmonics.*

| 1750 | 5th harmonic |
| 2450 | 7th harmonic |
| 3150 | 9th harmonic |

and so forth. (The square wave is particularly useful as a test signal for an oscilloscope, so it will be covered in detail in the next section of this chapter.)

A triangle wave, shown in Fig 5-13, has the same harmonic content as a square wave—that is, the fundamental frequency, and all odd harmonics, but no even harmonics. The difference between a square wave and a triangle wave is the relative amplitude of the harmonics. In a square wave, the harmonics are very strong; in a triangle wave, they are very weak and often can be considered negligible. Sometimes a triangle wave is used in place of a true sine wave, when some harmonic distortion (the harmonic content of the triangle wave) can be reasonably tolerated.

**Fig. 5-13**   *A triangle wave has the same harmonic content as a square wave, but at much lower amplitudes.*

## Types of distortion

When distortion occurs, the appearance of the displayed waveform on your oscilloscope's screen will be altered in some way. The effects of distortion are usually quite easily visible on an oscilloscope.

There are a number of different types of distortion. Some forms of distortion will show up better on certain waveforms than on others, which is why more than one waveform is used in electronics testing. Probably the two most generally useful types of waveforms for distortion testing with an oscilloscope are the sine wave and the square wave (discussed in the next section of this chapter).

Several possible reasons why a waveform might be distorted are:

- One or more harmonics (a specific frequency range) might be attenuated more (or boosted less) than the rest of the waveform.

- One or more harmonics (a specific frequency range) might be boosted more (or attenuated less) than the rest of the waveform.

- New harmonics or enharmonics are being added to the signal from some undesired source, such as RF pick-up, signal leakage, or clipping.

Figure 5-14 shows a sine wave with severe clipping on its positive peaks. Notice that the top of this waveform is flattened. Such clipping is usually an indication that the signal is too strong for the circuit being tested. The circuit in question doesn't have sufficient power to reproduce the peaks of the signal. Clipping may show up on the positive peaks, as shown in the figure, or on the negative peaks, or both. When a signal is clipped, it more closely resembles a square wave, implying that strong odd-order harmonics are being added to the signal.

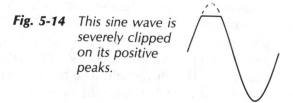

**Fig. 5-14**   *This sine wave is severely clipped on its positive peaks.*

In practical electronics work, most cases of clipping will be much less severe than this example. A particularly severe case of clipping is shown to make the effect as visible and obvious as possible.

Incidentally, when clipping occurs on a square wave, the only major effect is a reduction in the signal's amplitude. The clipped waveform is still a perfectly good square wave.

Figure 5-15 illustrates another common form of distortion: a sine wave being distorted by severe overshoot. This might be an indication of miscalibration somewhere in the circuit under test. A less common, but still possible, cause of this display signal could

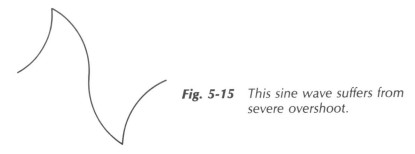

**Fig. 5-15**   *This sine wave suffers from severe overshoot.*

be miscalibration of the oscilloscope itself. That is, there is nothing wrong with the circuitry being monitored or its signals; the only problem is in faulty testing. Yet another possibility to consider is that the sine wave is picking up a spike or pulse signal of the same frequency from elsewhere in the circuit. If the spikes, or other distortion artifacts, show up at regular intervals, but not at the same point in each displayed circle, it means there is an interfering signal that does not have the same frequency as the desired signal.

Note that in some circuits, a displayed waveform like the one shown in Fig. 5-15 might not indicate any problem at all. Sometimes one signal rides on (or modulates) a second signal. It is always important to know what the monitored signal is supposed to look like at that point in the circuit before you draw any conclusions.

In waveforms with straight edges — such as sawtooth waves, triangle waves, and square waves — make sure that the straight edges are, in fact, straight. Any curvature is an indication of distortion in the signal. (This may or may not indicate an actual problem, depending on the purpose of the circuit being tested.)

Undesired curvature of the straight edges of a waveform might be the fault of the circuit being monitored, or it may be caused by the oscilloscope itself or its test probe. An uncalibrated test probe can easily cause this sort of problem.

## Isolating signal distortion

When you are attempting to isolate the source of any type of signal distortion, a dual-trace oscilloscope can be very helpful. (Refer to chapter 1 for more information on dual-trace oscilloscopes.) With this kind of test instrument, you can simultaneously monitor the input and output signals for any given circuit or subcircuit, making any differences very easy to see and compare.

It is usually best to set the oscilloscope's volts-per-division controls so that both signals are displayed at the same size (the same apparent amplitude). In other words, deliberately neutralize the effects of any overall signal amplification or attenuation occurring in the circuit stage. This way you won't be misled by differences in the overall signal amplitudes. It will be easier to spot minor differences between the two waveforms if they are the same basic size.

In some cases, it might be useful to bring the two (equal display amplitude) signals together on the oscilloscope's display screen by moving their baselines until they are overlapping. Any difference between the two signals often will stand out very noticeably under such circumstances.

Remember that some circuits are designed purposely to change the waveshape between the input and the output — that is, the two displayed signals aren't supposed to look alike. You must know what the circuit under test is supposed to be doing before you can draw any meaningful conclusions.

For example, switching circuits often deliberately clip non-square-input signals for cleaner and more reliable switching. Filter circuits, by definition, change the amplitude of some frequency components, but not others. This means the harmonic content of the test signal will be altered, which is bound to have an effect on the displayed waveform. Any modulation system combines two or more separate signals into a new, more complex signal.

There are many other possible examples of such deliberate distortion. If the output signal doesn't resemble the input signal, that doesn't necessarily mean there is any problem in the circuit being tested. The technician must know what he is looking for in each case when working with electronic circuits using an oscilloscope.

## Square-wave testing

A square-wave signal is more than a sine wave with the tops and bottoms clipped, although this is the way many commercial signal generators make them. Mathematically, a true square wave is a signal of periodic recurrence made up of an infinite number of odd harmonics of the fundamental frequency. You can make complicated scientific tests with square waves if you want, but there's one test you can do with no math involved.

If a good square wave, such as the one in Fig. 5-16, is fed into an amplifier and the output is a signal that looks anything like the original, it is a good amplifier. The first thing to do is check the frequency response of the scope by feeding the square-wave signal directly into the vertical amplifier. If your scope shows signs of tilt on the tops and bottoms of the square waves or any other troubles, you can switch the vertical amplifier off and feed the signal directly to the vertical-deflection plates. Most scopes have provisions for doing so. You can always have a good-sized signal at the output of an amplifier under test, so you'll get plenty of vertical deflection without using the amplification of the scope. Incidentally, a square-wave signal is also useful for finding horizontal nonlinearity and other troubles in the scope amplifiers.

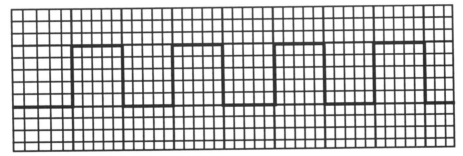

**Fig. 5-16**   *An oscilloscope display of a good square wave signal.*

In the waveform of Fig. 5-15, all you see are the tops and bottoms of the signal. The vertical lines aren't visible because they rise and fall too fast to register. This is the sign of a good square wave (and also a fairly good scope amplifier, even if it is just a service-type scope). However, even if the vertical lines are visible on your scope, it doesn't matter. What does matter is the distortion that might be added to the square-wave signal on its way through the amplifier being tested; all you need to do is compare the input and output signals to evaluate the amplifier's performance.

For instance, if the square wave at the output of the amplifier looks like Fig. 5-17, then the amplifier has poor low-frequency response. Poor high-frequency response shows just the opposite reaction; the tops of the waves slant up toward the right.

If you feed in a good square wave and get something out like Fig. 5-18, the amplifier is *differentiating* the square pulses (making spikes out of them), usually because of a load resistor that has gone down in value or an open capacitor, etc.

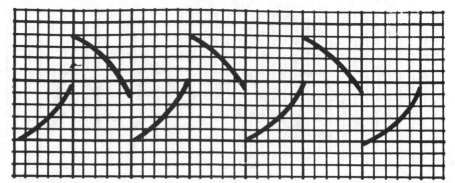

**Fig. 5-17**   *A circuit with poor low-frequency response distorts a square-wave signal like this.*

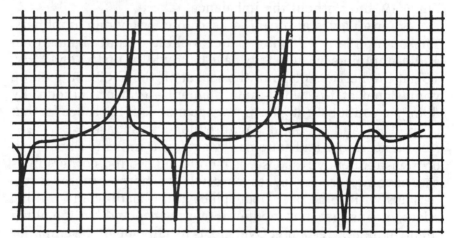

**Fig. 5-18**   *This scope trace reveals that the circuit being tested is differentiating the square-wave pulses.*

The type of distortion seen in Fig. 5-19 could be caused by a leaky coupling capacitor, a load resistor that has fallen off in value, incorrect bias, etc. This basic pattern also can be found in push-pull circuits where the two halves aren't balanced. Note that the top halves of the wave differ from the bottom. This situation always indicates an imbalance somewhere.

Don't be too critical. If your output waveform is a reasonable duplicate of the input, take it and be happy. An amplifier that could reproduce a perfect square wave in the output from a square-wave input of 1,000 Hz would need a bandwidth of more than 20,000 Hz. Even the best hi-fi amplifiers have a maximum bandwidth of only about 50,000 Hz. Cheaper amplifiers have a

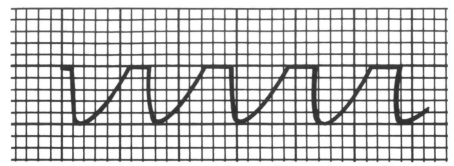

**Fig. 5-19**    *Several different circuit defects could cause this type of distortion.*

maximum bandwidth of about 18,000 Hz, so you'll never find perfect reproduction. In fact, if the output signal is as good as Fig. 5-19, but without the imbalance, take it.

Square waves also can be used to check video amplifiers. These amplifiers will show better looking output signals than audio amplifiers because they have bandwidths up to 3 or 4 MHz. To get the best results, pull the socket off the CRT and pick up your signal at the video-input element—grid or cathode. By doing so, you substitute the input capacitance of the scope for that of the CRT and get better high-frequency response.

Slanting tops and bottoms of waveforms indicate a poor low- or high-frequency response, as mentioned. Round corners indicate poor high-frequency response, while a sharp-cornered wave, perhaps with some overshoot spikes at the leading and trailing edges, means that the amplifier is overpeaked and has too much high-frequency response. Overshoots usually mean the amplifier circuits have a tendency to ring at high frequencies.

# Time and phase measurements

One way to describe the oscilloscope is to say that it measures and displays changes in a voltage over time. A dual-trace oscilloscope can be particularly useful in monitoring time and phase relationships between pairs of signals. Does the voltage change the way it should and when it should?

You cannot make true phase tests on a single-trace oscilloscope. The phase of a signal is meaningless unless it is in comparison with some other signal. You can't say "such-and-such a signal is in phase" unless you define what it is in phase with.

To compare the timing of two or more signals, you should generally avoid the "normal" triggering mode. In this mode, you should recall, the oscilloscope completely draws trace A, then it completely draws trace B. The two displayed signals are inevitably time-shifted from one another by one complete cycle of the oscilloscope's sweep frequency.

It is fairly easy to determine if a one-shot or infrequently occurring event occurs at the correct time with a dual-trace oscilloscope. For example, let's say a brief high-voltage spike should be generated 3 mS (0.003 second) after some other switching voltages goes from low to high. Assuming that the oscilloscope is set up for a time base of 1 mS/division, the displayed spike should show up exactly three divisions after the low-to-high transition in the switching voltage. Obviously, there is no way to directly check this correlation in timing with a single-trace oscilloscope. You need to be able to see both signals at once.

Often in an electronic circuit, two signals must be in phase with each other, or one of them must be phase-shifted by a specific amount. For simplicity, assume for now that both signals have the same frequency. When the two signals are in phase, they will begin and end their cycles at the same instant. Obviously, they must be at the same frequency (or harmonically related) to stay in phase with each other. Signals of differing frequencies will drift in and out of phase with one another.

Any cycle is made up of 360 degrees. If two signals of the same frequency are exactly 360 degrees out of phase with each other, then you might as well consider them in phase, since any one cycle of either signal is the same as the one before it or after it. Both signals begin and end a cycle in unison. So what if signal A is producing cycle 5, while signal B is on cycle 6?

A pair of in-phase signals are illustrated in Fig. 5-20. This same illustration would also apply for two signals that are 360 degrees (one complete cycle) or 720 degrees (two complete cycles) out of phase with one another.

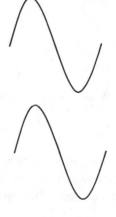

**Fig. 5-20**    *These sine wave signals are in phase.*

One-quarter cycle is 90 degrees. A 90-degree phase shift is shown in Fig. 5-21. Notice that (with sine waves), the second signal begins its cycle as the first signal reaches its positive peak.

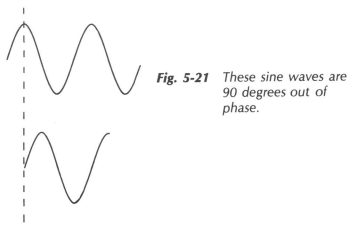

**Fig. 5-21**    *These sine waves are 90 degrees out of phase.*

One-half cycle is 180 degrees. The two signals shown in Fig. 5-22 are 180 degrees out of phase with one another. Assuming two signals have the same waveshape, frequency, and amplitude, but are 180 degrees out of phase with each other, as in the illustration, mixing these two signals together will cause them to cancel each other out. The result will be 0, or no signal at all. As one signal is going positive, the other signal is going negative by an identical amount, so the effective combined value is always 0 for each point throughout the cycle.

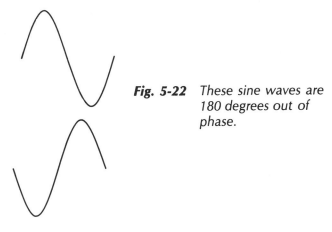

**Fig. 5-22**    *These sine waves are 180 degrees out of phase.*

In many electronics circuits and systems, it is necessary for the technician to know how much phase shift a given section of circuitry causes. This information can be determined easily with a dual-trace oscilloscope simply by comparing the phase of the input signal with that of the output signal from that circuit stage.

# Using a sweep-frequency signal generator

In working with an oscilloscope, a signal generator of some sort is often used as a predictable, known signal source. It would be hard, for example, to check a stereo system amplifier for distortion if the program signal was continuously varying (as in a musical recording). You would have considerable difficulty spotting distortion in the displayed waveforms because you don't know for sure what they are supposed to look like. Even comparing the input and output signals with a dual-trace oscilloscope wouldn't help very much because the signals are so complex and changing from instant to instant. It would be virtually impossible to make an accurate or reliable comparison under such circumstances.

A signal generator produces a regular, repeating simple waveform. All characteristics of this signal can be precisely known. As long as you leave the generator's controls alone, these characteristics will remain steady. Detailed comparisons can be made easily.

Now, if you want to change any parameter of the test signal for any reason, you must manually adjust the controls on the signal generator. In most test procedures, this level of control is precisely what is wanted. In some special cases, however, it might prove to be a time-consuming nuisance.

In checking the frequency response of a piece of equipment, you obviously need test signals running through the entire range of the equipment. With an ordinary signal generator, you would need to check the display and make notes, then readjust the frequency of the test signal, and repeat, usually dozens of times, to get anything close to an accurate picture of the equipment's true frequency response.

A better and more efficient way to perform such a test is to use a special type of signal generator that features sweep frequencies. In a sweep-frequency signal generator, a voltage-controlled oscillator (VCO) generates the actual test signal. A VCO automatically varies its frequency in response to a control voltage at its input. By changing the control voltage, you can change the signal frequency. Now, if you just use dc voltages, you have nothing special. The sweep-frequency signal generator will function just like an ordinary signal generator.

A sweep-frequency signal generator also has a second oscillator built into it, the *sweep oscillator*. It puts out a sawtooth or ramp waveform, as illustrated in Fig. 5-23. Usually the sweep oscillator is operated at a low frequency.

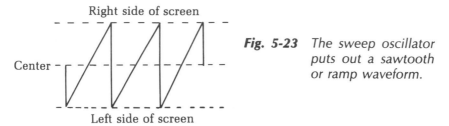

**Fig. 5-23** *The sweep oscillator puts out a sawtooth or ramp waveform.*

Because the waveshape is a ramp, the voltage starts at some minimum value, smoothly builds up to a maximum value, then drops quickly back down to the minimum value to start the next cycle. The output signal from the sweep oscillator is used as a control voltage for the VCO. At any given instant, the output frequency of the VCO is proportionate to the control voltage it presently sees at its input. As the sweep signal goes through its cycle, the signal generator's output frequency will start at some minimum value and smoothly sweep up to a maximum value, before dropping back down to the minimum frequency and starting over.

If you feed this swept signal through the equipment being tested and monitored on an oscilloscope, you can see a complete picture of the equipment's frequency response. Any peaks or dips in the frequency response will be clearly visible on the oscilloscope's display.

You can use a sweep-frequency signal generator to test the overall frequency response of amplifiers and other circuits, as well as to determine the cut-off frequency and roll-off slope of a filter. This technique also can help you to align and set the resonant frequency of a tuned circuit. (See chapter 9.)

# Lissajous figures

If the oscilloscope could perform only the functions described so far in this chapter, it would be an extremely useful piece of test equipment. The oscilloscope can do a lot more, however, making it truly invaluable for the electronics technician. There is an important alternate mode of operation for an oscilloscope that is even more powerful and versatile than the basic tests described so far: the use of Lissajous figures.

Don't let that complicated sounding name throw you. It has no hidden or deep meaning. Lissajous figures are simply named for an important scientist.

In the ordinary operation of an oscilloscope, the vertical (Y-axis) signal is generated within the oscilloscope itself. It is a sawtooth wave employed as a sweep signal that moves the electron beam from left to right across the face of the CRT screen at a specific, consistent rate. Although frequency measurements can be made in this mode, as described earlier in this chapter, such measurements are inevitably rather crude and not terribly accurate. It is difficult to estimate partial divisions when the test (horizontal, X-axis) signal is not an exact harmonic of the oscilloscope's sweep frequency. Also, phase comparisons require a more expensive dual-trace oscilloscope, and small phase shifts might be quite difficult to see. More accurate phase and frequency measurements can be made with even a single-trace oscilloscope by using Lissajous figures.

To create a Lissajous figure on an oscilloscope, two external signal sources are required. The oscilloscope's internal sweep-signal generator is not used in this mode of operation. Signal A is connected to the X-axis input, just as in any standard oscilloscope measurement. At the same time, signal B is connected to the oscilloscope's Y-axis input. With these connections, a closed-loop pattern of some sort will be drawn on the oscilloscope's CRT screen. This pattern is called a *Lissajous figure*.

The exact shape and size of the displayed Lissajous figure is determined by several factors of the two input signals: their waveshapes, their relative amplitudes, their relative frequencies, and their phase relationships. Notice that, for the most part, absolute quantities are not relevant, only the relative differences between the two input signals. Of course, the settings of the oscilloscope's various controls also will affect the size and placement of the Lissajous figure.

To keep things as simple as possible, assume that all input signals are in the form of sine waves. Harmonic content in one or both of the input signals will result in more complex Lissajous figures. Also assume that the input signals do not contain any dc offset voltages.

First, let's assume that the same sine wave signal is simultaneously applied to both the X-axis and Y-axis inputs. The resulting Lissajous figure will be a slanted line, as shown in Fig. 5-24. The length of this line reflects the amplitude of the input signal and the setting of the oscilloscope's volts-per-division control.

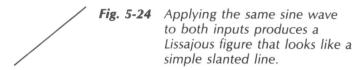

**Fig. 5-24** *Applying the same sine wave to both inputs produces a Lissajous figure that looks like a simple slanted line.*

Many people working in electronics, including many well-educated and experienced technicians seem to find Lissajous figures rather mysterious and hard to understand, so let's take a look at just how and why this slanted line is displayed on the oscilloscope's screen in some detail. When the instantaneous amplitude of the signal voltage is 0 V (the baseline value), the oscilloscope's electron beam will strike the exact center of the CRT screen. At this point, there is no signal to deflect the electron beam in either the horizontal or the vertical dimensions.

During the first 90 degrees of the cycle, the signal voltage increases smoothly up to its peak voltage. The electron beam is moved upward because of the increasing X-axis voltage. This same positive-going voltage is simultaneously applied to the Y-axis, so the electron beam is also deflected to the right. In other words, the top half of the slanted line is drawn.

From the 90-degrees point to the 180-degrees point of the cycle, the signal voltage drops back down from its peak value to 0. The reverse of what happened in the first quarter cycle occurs, so the same line is redrawn, albeit in the opposite direction. This makes no visible change in the displayed pattern. At the 180-egree point, the instantaneous signal voltage is once again 0 V, and the electron beam is aimed at the exact center of the CRT screen once more.

Now, during the next quarter cycle (from 180 degrees to 270 degrees), the voltage goes increasingly negative, until it reaches the negative peak value. The electron beam is forced to move downward and to the left, in a mirror image of what happened during the first quarter cycle. Drawing the bottom half of the Lissajous figure's slanted line.

Finally, from 270 degrees to 360 degrees, the signal voltage increases from the negative peak voltage back up to 0, redrawing the bottom half of the slanted line. At 360 degrees, the signal is once again 0 V, aiming the electron beam at the center of the CRT screen. This is also the 0-degree point for the next cycle. The complete Lissajous figure has been drawn, and in this case it is simply a straight line at an angle of 45 degrees. The 45-degree angle is dictated by the fact that the X- and Y-axis signals are completely identical.

What happens if the X- and Y-axis input signals are sine

waves of equal amplitude and frequency, but are out of phase with one another? As the phase shift increases from 0 (in phase), the Lissajous figure will start to open up into a slanted oval, as shown in Fig. 5-25. When the phase shift reaches 90 degrees, a perfect circle will be displayed, as illustrated in Fig. 5-26.

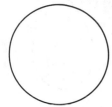

**Fig. 5-25**  *As the phase shift between the input signals increases from 0, the Lissajous figure starts to open up into a slanted oval.*

**Fig. 5-26**  *Two same-frequency sine waves at the inputs with a 90-degree phase difference produces a circle as the Lissajous figure.*

As the phase shift is increased beyond 90 degrees, the sides of the circle start to come together again, forming an increasingly narrow oval shape, until at 180 degrees, there again is a straight line (completely closed loop). This time, however, the line is slanted in the opposite direction as the in-phase display.

With phase shifts between 180 degrees and 270 degrees, the slanted oval opens up again, reaching a peak circle at the 270-degree point. For phase shifts of 270 degrees to 360 degrees, the oval closes up again, returning to the original closed, slanted line pattern it started out with. Remember, when two equal-frequency waveforms are 360 degrees out of phase with one another, there is no practical difference from the in-phase condition. In effect, 360 degrees of phase shift equals 0 degrees of phase shift.

For the next example, again assume the two sine wave input signals have equal amplitudes, but this time, let's say the vertical signal frequency is twice that of the horizontal signal. For each complete vertical cycle, only one-half the horizontal cycle is displayed on the oscilloscope's screen, appearing as a closed loop. The second (negative) half of the horizontal signal's cycle combines with a second complete vertical signal cycle, producing a second closed loop in the Lissajous figure displayed by the oscilloscope. In other words, a figure-eight pattern is produced as shown in Fig. 5-27.

When the vertical signal frequency is only half the horizontal signal frequency, there is pretty much the same effect. It takes two

**Fig. 5-27**  *If the vertical frequency is exactly twice the horizontal frequency, the Lissajous figure will appear as a figure-8.*

complete horizontal cycles to draw a complete vertical cycle, so the Lissajous figure in this case is a sideways figure-eight, as illustrated in Fig. 5-28.

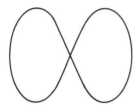

**Fig. 5-28**  *If the horizontal frequency is exactly twice the vertical frequency, the Lissajous figure will appear as a sideways figure-8.*

Assuming the two input signals are harmonically related, the number of closed loops in the displayed Lissajous figure indicates the frequency relationship of two signals. For example, Fig. 5-29 shows the Lissajous figure for a horizontal signal with a frequency three times that of the vertical signal. If the horizontal signal frequency is four times the vertical signal frequency, the Lissajous figure shown in Fig. 5-30 is displayed. If the vertical frequency is higher than the horizontal frequency, these patterns will be turned on their sides, as shown in Figs. 5-31 and 5-32.

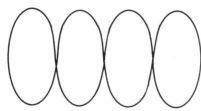

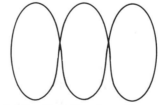

**Fig. 5-29**  *This Lissajous figure indicates the horizontal frequency is three times the vertical frequency.*

**Fig. 5-30**  *This Lissajous figure indicates the horizontal frequency is four times the vertical frequency.*

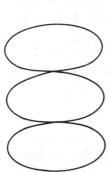

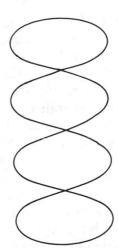

**Fig. 5-31**  *This Lissajous figure indicates the horizontal frequency is one-third the vertical frequency.*

**Fig. 5-32**  *This Lissajous figure indicates the horizontal frequency is one-fourth the vertical frequency.*

If you know the frequency of one of the signals, you can calculate the second signal frequency just by counting the number of loops in the Lissajous figure displayed on your oscilloscope's screen.

The standard 60 Hz line frequency often is used as the reference signal for relatively low-frequency Lissajous figure tests because this frequency is held to very exacting standards. If the 60 Hz line signal is used as the vertical signal and the Lissajous figure display shows three loops that are stacked horizontally, you know that the horizontal signal frequency must be equal to:

$$\text{horizontal frequency} = \text{no. of loops} \times \text{vertical frequency}$$
$$= 3 \times 60$$
$$= 180 \text{ Hz}$$

On the other hand, if the same test setup with a different unknown signal also produces a display of three complete loops, but this time they are arranged vertically instead of horizontally, then the horizontal frequency must be equal to:

$$\text{horizontal frequency} = \frac{\text{vertical frequency}}{\text{no. of loops}}$$
$$= \frac{60}{3}$$
$$= 20 \text{ Hz}$$

This is all very simple and straightforward. In the real world, however, electronic signals are very rarely cooperative enough to be exact whole number multiples of a convenient reference frequency. Fortunately, Lissajous figures also can be used to determine nonharmonic-frequency relationships. The process is just a little more complicated.

As the frequency relationship between the horizontal and vertical signals grows more complex, so does the resulting Lissajous figure. Specifically, the Lissajous figure grows *nodes*, or additional loops.

With the simple cases described so far, the number of nodes along at least one axis has always been one. A complex Lissajous figure can have two or more nodes along either or both the horizontal axis and the vertical axis.

For example, the Lissajous figure shown in Fig. 5-33 has five horizontal nodes and three vertical nodes. They are usually expressed in the form of a ratio. In this case, of course, the ratio is 3 : 5. Assuming that you know the vertical signal's frequency (the reference signal) and want to find the horizontal signal's frequency, you divide the reference (vertical) frequency by the number of horizontal nodes, and then multiply by the number of vertical nodes. That is:

$$F_h = VN \times \frac{F_v}{HN}$$

where $F_h$ is the horizontal frequency, $F_v$ is the vertical frequency, VN is the number of vertical nodes, and HN is the number of horizontal nodes.

For this example, once more assume that the reference (vertical) signal is the 60 Hz line frequency. If the Lissajous figure of Fig. 5-33 is displayed, the horizontal frequency must be equal to:

$$F_h = 3 \times \frac{60}{5}$$
$$= 3 \times 12$$
$$= 36 \text{ Hz}$$

A 3 : 2 Lissajous figure is shown in Fig. 5-34. Figure 5-35 shows a Lissajous figure with a 5 : 2 ratio. Of course, many other combinations are also possible.

When working with complex Lissajous figures, you must be very careful and precise. It is all too easy to miscount the nodes in very complex Lissajous figures. You should also be aware that

**Fig. 5-33** This Lissajous figure has five horizontal nodes and three vertical nodes.

**Fig. 5-34** This Lissajous figure has a 3 : 2 ratio.

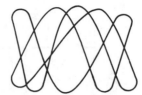

**Fig. 5-35** If the signal frequency ratio is 5 : 2, this Lissajous figure will be displayed.

sometimes certain phase relationships between the horizontal and vertical signals can obscure certain lines in the display, which are overlapping and therefore cannot be seen as separate lines or nodes.

As a general rule of thumb, a Lissajous figure is easiest to read when the two input signals have a phase difference of 90 degrees. Many modern oscilloscopes have a phase-adjustment control, which can be used to get a clearer Lissajous figure. You don't even need to know the actual phase relationship between the two signals. Just turn the oscilloscope's phase adjustment control until you get the clearest and easiest to read Lissajous figure display.

If, in a given test, the displayed Lissajous figure is excessively complex—a half dozen or more nodes along either axis—try changing the reference frequency. Often you will get a simpler, easier to read pattern. In some cases, minor frequency differences between the two input signals can cause the displayed Lissajous figure to rotate slowly.

Lissajous figures also can be used to test a wide variety of electronic components. Figure 5-36 shows a simple circuit that can be used for this test. A suggested parts list for this circuit appears in Table 5-2. Select the value of current-limiting resistor R1 to suit the neon lamp you are using. Some neon lamps come with a built-in current-limiting resistor, allowing you to eliminate R1 as a separate component in the circuit.

Notice that this project uses ac line current, so there can be serious shock or fire risks if you are not careful. Make sure all conductors are fully insulated. Do not omit the fuse or substitute a fuse with a higher rating than the one suggested in the parts list. Make connections to this circuit only when the power is disconnected. Ideally, you should unplug the circuit, but at the very least, turn off the power switch (S1). Please do not take foolish

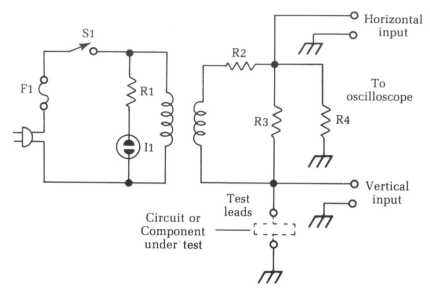

**Fig. 5-36**  *This circuit creates a useful Lissajous figure to test many different types of electronic components.*

**Table 5-2   The suggested parts list for the Lissajous figure component tester circuit of Fig. 5-36.**

| | |
|---|---|
| T1 | 6.3 Vac power transformer |
| I1 | 120 V neon lamp |
| R1 | see text |
| R2 | 330 Ω 1/2 watt 10% resistor |
| R3 | 150 Ω 1/2 watt 10% resistor |
| R4 | 1K 1/2 watt 10% resistor |
| F1 | 1/16 amp fuse and holder |
| | AC plug |
| | test clips |
| | binding posts (for connections to oscilloscope) |

chances. Cutting corners to save either time or cost is definitely not worth the risk.

This circuit can test components either in circuit or out of circuit. Do not apply power to the circuit being tested while this unit is hooked up. You could do some serious damage.

If the oscilloscope just displays a flat horizontal line, there is a short or a very low resistance between the test terminals. Unless the component in question is a very low valued resistor, it is probably defective. Similarly, a flat vertical line usually indicates that the component being tested is open and should be replaced, although high-valued resistors also might give this sort of display.

If this circuit's test leads are connected across a resistor, the displayed Lissajous figure will be a straight, slanted line, as illustrated in Fig. 5-37. The angle of this line will indicate the approximate resistance. A fully horizontal line indicates a resistance of about 10 $\Omega$ or less. A fully vertical line is displayed if the measured resistor has a value of about 100 $\Omega$K (100,000 $\Omega$). As with any in-circuit resistance test, any parallel resistances in the circuit might affect the reading. Be careful.

A potentiometer can be tested in pretty much the same way as an ordinary fixed resistor. After all, a potentiometer is just a manually variable resistor. If you connect the test leads across the potentiometer's two outer terminals, it will act just like a fixed resistor with the maximum resistance value of the potentiometer. The setting of the control shaft is irrelevant.

For an even more useful test, connect one test lead to the potentiometer's center terminal and the other test lead to either of the potentiometer's end terminals. Turn the control shaft slowly and as smoothly as possible through its entire range. The displayed line should move through its range of angles smoothly and evenly. If the display jumps about as you turn the potentiometer's control shaft, the component has problems. Try cleaning it. If cleaning doesn't help, the resistive element is probably bad, and the potentiometer should be replaced. (See chapter 6 for more information on testing and cleaning potentiometers.)

If the test leads of this circuit are connected across a capacitor, you should get a slanted oval, as illustrated in Fig. 5-38. Using the component values suggested in the parts list, the circuit can test circuits over a range extending from about 0.05 $\mu$F up to about 20 $\mu$F. To test smaller valued capacitors, increase the value of resistor R4. Similarly, decrease the value of this resistor to test large, electrolytic-type capacitors.

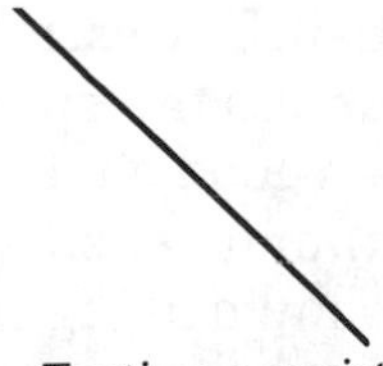

**Fig. 5-37**  *Testing a resistor with the circuit of Fig. 5-36 should produce a straight, slanted line for the Lissajous figure.*

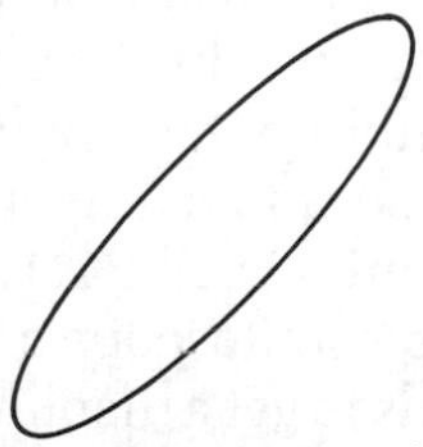

**Fig. 5-38**  *This is the expected Lissajous figure when testing a capacitor with the circuit of Fig. 5-36.*

An inductor will give a similar display to a capacitor; however, as illustrated in Fig. 5-39, the oval will slant in the opposite direction because capacitance and inductance cause opposing phase shifts. With the component values suggested in the parts list, the range of measurable inductances runs from approximately 300 mH up to about 2 H, which should be a more than sufficient range for virtually all practical electronics work. Because the windings of a transformer are coils, they can be tested in the same way as inductors.

A semiconductor junction creates an interesting Lissajous figure with this test circuit. A typical display is shown in Fig. 5-40. If you get just a straight horizontal line or a straight vertical line, with no bend, then the semiconductor junction being tested is defective. It is probably shorted or open. When making in-circuit tests, however, watch out for possible parallel resistances, which could be confusing the tester's results. Note that this test will work for diodes or any two leads of a biopolar transistor or JFET.

**Fig. 5-39** *This is the expected Lissajous figure when testing an inductor with the circuit of Fig. 5-36.*

**Fig. 5-40** *A semiconductor junction should produce a Lissajous figure that looks like a sharply bent line, using the circuit of Fig. 5-36.*

For best results, calibrate any type of test equipment before using it. This test circuit itself has no calibration, but you should calibrate the oscilloscope you are using with it. The easiest way to do so is to connect a known good resistor across the circuit's test terminals. A 1K (1,000 Ω) resistor would be a good choice. Adjust the oscilloscope's trace positioning (offset) and sensitivity controls to get a display that looks like the one shown in Fig. 5-37. The displayed line should slope from the lower left to the upper right as shown in the diagram. That's all there is to the calibration procedure for this versatile test setup.

# Reactance and impedance measurements

Ordinary dc resistance is easy to measure with an ohmmeter, but in ac circuits, resistance becomes much more complex. When ac signals are used, the resistance is not a simple, constant value parameter. Instead, the ac resistance, or impedance, changes with signal frequency and phase.

We will get to reactance and impedance measurements with an oscilloscope shortly, but first we need to run through a little background theory. This will presumedly be a review for an electronic technician, so we won't go into much depth or detail here.

Impedance is made up of dc resistance, plus two types of ac (frequency-dependent) resistance called *reactances*. A pure capacitor exhibits capacitive reactance, and a pure inductor exhibits inductive reactance. The capacitive reactance in any circuit is always 180 degrees out of phase with the inductive reactance. For this reason, the two types of reactance cannot be simply added together.

The total impedance of a capacitor is the sum of its dc resistance and the capacitive reactance. The dc resistance is a constant for any given capacitor, regardless of the signal frequency. The capacitive reactance, however, is determined by both the component's capacitance value and the applied frequency. The formula for capacitive reactance is:

$$X_c = \frac{1}{2\pi\ FC}$$

where $F$ is the signal frequency in Hertz, $C$ is the capacitance in farads — do not use microfarads ($\mu$F) or picofarads (pF) in this equation — and $X_c$ is the capacitive reactance in ohms. The symbol $\pi$ is a Greek letter pi. It represents a mathematical constant with a value of approximately 3.14. Since the value of $\pi$ never changes, we can rewrite this equation as:

$$X_c = \frac{1}{6.28\ FC}$$

Assuming that the capacitance holds a constant value (which is usually the case for most electronic components), as the applied frequency increases, the capacitive reactance decreases, and vice versa.

A coil or an inductor exhibits a dc resistance (usually fairly small) and an inductive reactance. Once again, the dc resistance is a constant value, regardless of the applied signal frequency, while the inductive reactance is determined by the inductance of the coil and the applied frequency, according to this formula:

$$X_1 = 2\pi \; FL$$

Where $F$ is the signal frequency in Hertz, $L$ is the inductance in henries — do not use millihenries (mH) or microhenries (uH) in this equation — and $X_1$ is the inductive reactance in ohms. Again, since the value of pi never changes, we can rewrite this equation as:

$$X_1 = 6.28FL$$

Unlike capacitive reactance, inductive reactance increases as the applied frequency is increased, and vice versa. This difference is a result of the opposing phase shifts of capacitive reactance and inductive reactance.

Because of the phase shifts, the three elements making up impedance cannot simply be added together. The total ac resistance or impedance is equal to:

$$Z = \sqrt{R^2 + (X_1 - X_c)^2}$$

where $Z$ is the impedance, $R$ is the dc resistance, $X_1$ is the inductive reactance, and $X_c$ is the capacitive reactance. All of these values are in ohms.

Both capacitive reactance and inductive reactance are frequency-dependent. Capacitive reactance decreases as the applied frequency increases, while inductive reactance increases with any increase in the signal frequency. At some specific frequency, the capacitive reactance will exactly equal inductive reactance, so these two values will cancel each other out. This leaves an impedance of:

$$\begin{aligned} Z &= \sqrt{R^2 + O^2} \\ &= \sqrt{R^2 + O} \\ &= \sqrt{R^2} \\ &= R \end{aligned}$$

Notice that this is the lowest possible impedance this particular circuit can ever exhibit. The point at which this happens is known as *resonance*. The capacitive reactance is exactly equal to the inductive reactance at the resonant frequency.

Now, how can you put an oscilloscope to work in practical reactance and impedance measurements? To monitor an ac resistance (impedance or reactance), connect the voltage waveform to the oscilloscope's vertical input and the current waveform to the horizontal input.

A typical test setup is shown in Fig. 5-41. Notice that the oscilloscope's internal sweep generator is not used in this test. You are applying ac signals to both the vertical and horizontal inputs, so the result will be a Lissajous figure, as discussed previously.

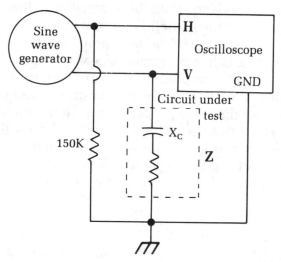

**Fig. 5-41**   *This is the test setup to monitor an ac resistance with an oscilloscope.*

The relative values of the resistance (R), reactance (X) and impedance (Z) will appear in the displayed Lissajous figure's proportions. The Lissajous figure displayed should be a single closed loop, since the vertical and horizontal input signals, by definition, have identical frequencies. The displayed loop will usually be an oval, not a circle, and it usually will be displayed at an angle.

Make sure you first calibrate the oscilloscope so that the displayed pattern is centered on the CRT screen. That is, when no external signal is applied to either input, you should see just a single dot in the center of the screen. Now apply the test signal and look at the resulting Lissajous figure.

Count the number of divisions covered by the Lissajous figure along the 0 line. This value represents the measured reactance. If the loop is slanted to the right, the overall reactance is primarily capacitive. If the loop is slanted to the left, then the overall reactance is primarily inductive. If there is no slant, then the signal frequency being used is the resonant frequency for the circuit being tested. If there is no reactive component, the circuit has pure dc resistance only.

The next step is to count the number of divisions between the opposite peaks of the displayed loop. This value represents the total impedance (Z). Obviously Z should always be greater than X (reactance). If the X value appears to be larger, then some mistake has been made in the test setup. If the reactance really was greater than the total impedance, then the dc resistance (R) would have to have a negative value, which is not possible.

You now can find the R value by graphing the X and Z values. Draw a downward line on graph paper corresponding to the X value. Then, with a compass, draw an arc that has a radius equal to the Z value, starting from the bottom-most point of the X line. Now, draw a straight line perpendicular to the top of the X line until it intersects the Z arc. The length of this line will correspond to the value of R, the dc resistance. This will all be easiest to do on graph paper, but regular paper and a ruler can be used. Figure 5-42 shows a typical triangle drawn by this method.

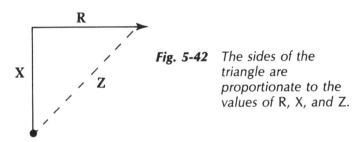

**Fig. 5-42**  *The sides of the triangle are proportionate to the values of R, X, and Z.*

Notice that there are no definite units in these measurements. You don't know the ohms values of R, X, or Z. The oscilloscope won't do that for you. To find the actual values in ohms, measure the dc resistance with your ohmmeter. You then will be able to find the proportional values of X and Z.

Let's try a quick example. Let's say the Lissajous figure gave you 9 units for the Z value, and 3.5 units for the X value. It doesn't really matter just what these "units" are, as long as they are consistent. When you graph these values as just described, you find that the R line is about 8.3 units long.

Measuring the dc resistance with an ohmmeter, you get a value of 6,400 $\Omega$ (6.4K). This means 8.3 "units" equals 6,400, so you can find out how much each unit is:

$$6,400 = 8.3 \times \text{UNIT}$$
$$\text{UNIT} = \frac{6400}{8.3}$$
$$= 771 \ \Omega$$

Since you now know the ohms value of one unit (771 $\Omega$), you can find the comparable ohms values for the $X$ and $X$ values:

$$X = 3.5 \times \text{UNIT}$$
$$= 3.5 \times 771$$
$$= 2,698 \ \Omega$$

$$Z = 9 \times \text{UNIT}$$
$$= 9 \times 771$$
$$= 6,939 \ \Omega$$

Notice that if $Z$ is less than either $R$ or $X$, or more than their sum $(R + X)$, you did something wrong somewhere along the line.

This procedure won't give exactly precise ohm values for the reactance or impedance because there almost inevitably will be some rounding off of values during the equations, and measuring precise units on an oscilloscope is difficult. Fortunately, these cumulative errors will generally be negligible for most practical electronics work. Also, since a better, more direct method of reactance or impedance measurement usually isn't available to the average electronics technician, this method works very well indeed.

## Testing for filter ripple with the scope

One of the best and fastest ways to check the condition of the filter capacitors in all ac power supplies (tube or transistor) is to touch the scope probe to the rectified output — to the V+ point that supplies the operating dc voltages to the circuits. The voltage at this point should be a pure dc, which would make a straight line on the scope, but normally there's a small amount of the ac left. This ac is called *ripple*; it looks like Fig. 5-43.

Note the generally triangular shape of the waves, and also note that you see a high peak, then a lower one, then another high one, etc. This unevenness is a result of the fact that this is a 120 Hz waveform in a full-wave rectifier circuit. In TV sets, the vertical output circuit takes a heavy pulse of current at a 60 Hz frequency, which causes the drop in voltage on every second peak of the ripple.

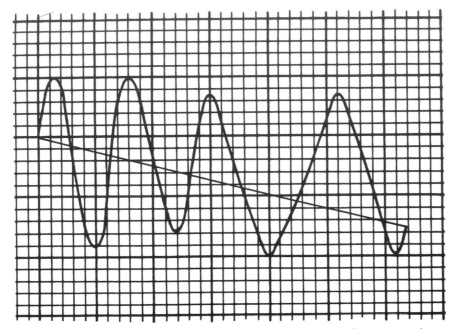

**Fig. 5-43**   *This oscilloscope display reveals ripple in the ac signal.*

You might have encountered TVs where a dark (or light) bar floats up and down through the picture. This trouble is caused by insufficient filtering in the V+ circuits or by faulty silicon rectifiers. If you check the power-supply ripple, you'll see a big difference between the high and low peaks, and one usually will change phase with respect to the other. This ripple waveform will writhe slowly if you lock the scope sweep to the local line frequency, as illustrated in Fig. 5-44.

The cure is to provide more filter capacitance, since ripple is caused by a lack of capacitance in the filter. Either the circuit didn't have enough to start with, or one of the original capacitors has gone down in value. If the original capacitors seem to be o.k., add more capacitance until the ripple is reduced and the bar vanishes.

The normal peak-to-peak value of the ripple will be given on the schematics of recent sets. As an average, it shouldn't be more than about 2 V peak to peak at the filter output. On the rectifier output (filter input), it will be higher—usually somewhere around 10 V 12 V p-p, or even more. However, the ripple at the filter output is the one that causes trouble, so check it first.

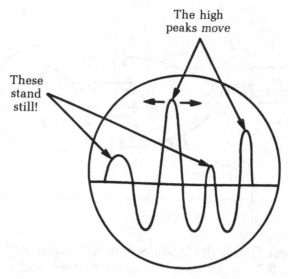

The high peaks *move*

These stand still!

**Fig. 5-44**  *The ripple waveform will writhe slowly if the oscilloscope's sweep is locked to the local line frequency.*

# Checking CB transmitter modulation with the scope

The oscilloscope will do a good job of measuring the audio modulation of a CB transmitter if you use it correctly. Measure the RF output with and without modulation. However, even the wideband color scopes with bandpass up to 5 MHz won't do too much on a 27 MHz RF signal. You'll get only a line on the scope screen. When you modulate the transmitter, the modulation will appear on the screen. This is what happens: The vertical amplifier in the scope simply won't pass the high-frequency RF carrier, but it will detect the audio modulation and show a good-sized deflection.

To get a true picture of the RF output, feed it directly into the vertical plates of the scope CRT. This gives you an almost unlimited bandpass, but no gain, because you're going around the vertical amplifiers of the scope. You'll probably see a line about ½ inch in height at most, for the unmodulated carrier. If you feed an audio signal into the microphone (or whistle into it), you'll see the modulation. The hook-up for this test is illustrated in Fig. 5-45.

By estimating the increase in pattern height, you can tell if the transmitter is modulating properly. If it should be overmodulating, the carrier will break up into the characteristic string-of-

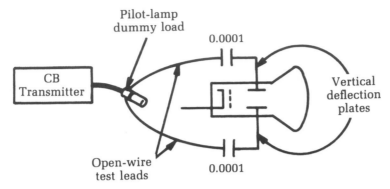

**Fig. 5-45**    *An oscilloscope can be used to check a CB transmitter's modulation with this hook-up.*

beads pattern. Check the modulator or RF stages; too much audio or too little RF output have the same pattern and effects.

Another good cross-check for modulation is to use a pilot lamp as a dummy antenna. With the transmitter unmodulated, it should glow a bright yellow. Whistling into the microphone should make it glow more brightly.

Some older scopes have provision for connecting directly to the vertical deflection plates by turning a switch on the front marked AMP OUT. Others have terminal boards on the back of the scope case with links that can be opened to give access to the vertical plates.

In either case, use plain, unshielded test leads for this test. Don't use shielded wires: the shunt capacitance of even the lowest capacitance coax will reduce the signal strength greatly, and there isn't any to spare. Put a small blocking capacitor in series with each lead—a .0001 $\mu$F (100 pF). Connect the test leads directly across the dummy-load lamp.

If your scope uses only one vertical-deflection plate with the other tied internally to one of the horizontal-deflection plates, use the free plate as the hot lead and connect the common plate to the CB-set chassis. Keep the test leads well apart and make sure they aren't moved during the test. Movement could change the shunt capacitance and the readings.

# 6
# Component tests

THIS CHAPTER DEALS WITH THE PROBLEMS AND SOLUTIONS OF DEALING with discrete components.

## Detecting thermal drift in resistors

Thermal drift in resistors is one of the most common and annoying troubles in all kinds of electronic equipment. It causes such symptoms as "it plays for an hour and then acts up." If a carbon resistor changes in value as it heats up, it changes the circuit characteristics.

The time constant will tell you a lot. If the trouble shows up inside of 15 minutes, it is probably a result of *self-generated heat* in a resistor, heat generated in the resistor because of the current it is carrying. Heat occurs in plate-load circuits, voltage-dropping resistors, etc. If the trouble takes an hour or so to appear, the resistor is being affected by heat traveling through the metal chassis or from a hot component close by.

The best test for a suspected resistor is to heat it up artificially and watch for the trouble to appear. For example, if you had a long-time-constant sync trouble, you could turn the set on, adjust it for correct operation, and then apply heat to each of the resistors in the sync circuit. Place the tip of a soldering iron on the body of each resistor, hold it there for 45 to 60 seconds, and watch for the sync trouble on the screen. Normal operating temperature in a TV set is 120° to 130°F. The tip of a soldering iron runs about 600°F, so don't hold it on the resistor too long—just long enough to get the resistor warmer than normal or too hot to hold a fingertip on it.

If the resistor has a tendency toward thermal drift, this test will reveal it.

The high-value (6, 8, 10, and 20 M) resistors found in age circuits are frequent offenders. You can find the guilty resistor by turning on the set, heating up each resistor in turn, and watching the screen for any sign of the original trouble. There might be more than one defective resistor in a given circuit, so check them all.

# Checking and repairing potentiometers

A *potentiometer* is simply a manually controlled resistor. The term is often informally shortened to *pot*.

The effective resistance of a potentiometer is determined by the position of a rotating shaft. Figure 6-1 illustrates the internal construction of a typical potentiometer. The two end terminals are connected to either end of a U-shaped resistive element. The control shaft is connected to a sliding element. Rotating the shaft moves the slider in one direction or the other along the length of the resistive element, creating a changing resistance. Let's assume that the slider starts off as close as possible to end terminal A and is slowly moved toward end terminal B. The resistance between the center terminal and end terminal A increases, while the resistance between the center terminal and terminal B decreases by a like amount. The resistance element can be designed so that the resistance changes linearly (Fig. 6-2) or logarithmically (Fig. 6-3), depending on the requirements of the specific application.

A trimpot, or trimmer potentiometer works in much the same way, except the control shaft can be replaced with a screw-

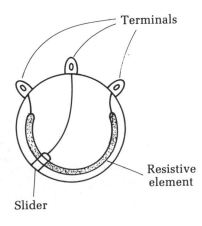

Terminals

Slider

Resistive element

**Fig. 6-1** *Inside a potentiometer, a slider is moved over a resistive element.*

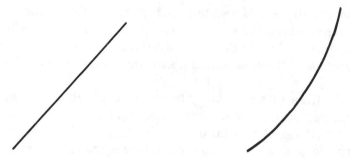

**Fig. 6-2**  *Some potentiometers have a linear response.*     **Fig. 6-3**  *Other potentiometers have a logarithmic response.*

driver adjustment. This adjustment reduces the size of the component and is more secure for "set-and-forget" calibration controls.

You occasionally might encounter a slide pot. This component is basically the same as a standard potentiometer, except the resistive element's normal U-shape has been straightened out into a line. The control shaft does not rotate; instead, it is moved back and forth over the straight-line resistive element. A slide pot is used in applications in which the position of the control shaft is to be directly visible, instead of indicated by markings on a dial. For example, slide pots are widely used in audio mixer consoles and graphic equalizers.

Some potentiometers, especially trimpots, have just two terminals, instead of the more common three just described. In such cases, there simply is no connection with one of the ends of the resistive element—there is just the center terminal and one of the end terminals.

A three-terminal potentiometer can easily be used in any application calling for a two-terminal potentiometer. Simply leave one of the end terminals disconnected and unused in the circuit. Some technicians prefer to short the unused end terminal to the center terminal. This method won't affect the operation of the potentiometer in any noticeable way.

## Testing potentiometers

Testing a potentiometer is usually fairly simple. You can use the ohmmeter section of any multimeter. It doesn't much matter what type of multimeter is used, as long as it is reasonably accurate in its coverage of the required resistance range.

As with any test of resistances, testing a potentiometer in circuit can give inaccurate readings because of parallel resistances in the circuit. When in doubt, temporarily break the connection between one of the test terminals of the component and the circuitry.

On a three-terminal potentiometer, first measure the resistance between the two end terminals. You should get the potentiometer's full rated resistance. The setting of the control shaft should have no effect at all on the resistance read between the two end terminals. For example, if you are testing a 10K potentiometer, you should get a reading of 10,000 Ω, more or less. Don't be too concerned about the exact resistance value. Most potentiometers are not precision components when it comes to their full-scale resistance, nor do they need to be in the vast majority of practical applications. As long as the resistance reading you get is in the right ballpark, you can assume that the potentiometer has passed this test.

The more important test is to monitor the resistance from the center terminal to one of the end terminals. It usually doesn't matter which one you use. (Of course, for a two-terminal potentiometer, there is no choice, since there is only one end terminal.) If possible, clip the test leads of your ohmmeter in place on the potentiometer's terminals to leave your hands free. Slowly and smoothly rotate the control shaft while watching the ohmmeter's pointer or display. As the control shaft is rotated toward the end terminal being used in the test, the resistance should decrease. Rotating the shaft away from the monitored end terminals should cause an increase in the resistance.

Rotate the control shaft as smoothly as possible. The resistance reading also should change smoothly and evenly. (Remember that some potentiometers change their resistance in a linear fashion, while others have a logarithmic response.) If the resistance jumps around, jiggles up and down, makes large skips, or gets "stuck" at a specific value over a few degrees of the shaft's rotation, a problem is indicated. Such erratic readings generally indicate that the resistive element is dirty or pitted, resulting in unreliable operation in the circuit. In some cases, there might be static or a crunching noise in audio circuits as the potentiometer's control shaft is rotated.

## Cleaning potentiometers

Slide pots are particularly prone to getting dirty because their construction inevitably leaves the resistive element more exposed to external contamination than in most standard round pots. To cure this problem, spray in some cleaner/lubrication spray, which is sold by almost any electronics parts store. It is often called "tuner cleaner." It is an aerosol spray can, with a small straw that can be connected to the nozzle to finely direct the spray. Direct the spray into any openings in the body of the potentiometer and then work the control shaft back and forth over its entire range several times to spread the cleaning fluid over the entire resistive element. It is a good idea to wait a couple of minutes to let the fluid dry, but this is rarely absolutely necessary. For a slide pot, you can easily insert the spray in the slot for the control slider's back and forth path.

When in doubt, clean all potentiometers, just to be on the safe side. Unless you use a ridiculous amount of spray, doing so will never hurt the circuit and even if dirt isn't causing the current symptoms, it is still likely to clean up dirt that hasn't built up to a noticeable problem yet. Cleaning the potentiometers is a form of preventive maintenance.

A lubricant in the cleaning spray is usually helpful, but a few components might be damaged by lubrication. When in doubt, use an unlubricated cleaning spray, even though it won't always be quite as effective.

If cleaning the potentiometer does not help, the resistive element might be badly pitted and permanently damaged. Try substituting a potentiometer that is known to be good. If the problem clears up, discard the original potentiometer as defective, and solder in a permanent replacement.

Many potentiometers, especially those used as volume controls, come equipped with a switch mounted on the back of the component. When the control shaft is rotated past a specific spot (usually near one end of its range of movement), the switch is activated. Sometimes these potentiometer switches develop problems. They might not make proper contact, or they might get stuck in one position or the other, ignoring the rotation of the potentiometer's control shaft.

Visually inspect the switch mechanism. Sometimes you might see that something is slightly bent. Use a pair of small needle-nose pliers to move it back into place. Be very careful not to do more damage.

Cleaner/lubrication spray will often solve poor switch contact problems. Spray some of the cleaning/lubrication solution into the switch mechanism and turn it on and off several times. If this doesn't work, you will probably have to replace the switch, which often entails replacing the entire potentiometer unit.

Stuck switches are usually unrepairable. The damage is probably permanent. It won't hurt to try to bend some small part back into place within the switching mechanism (if accessible), or even to try cleaning the switch unit, but don't get your hopes up too high, or spend a lot of time at it. You will probably end up having to replace the defective switch, which often means replacing the entire switched potentiometer unit.

# Capacitance testing

There are only two things you need to know about almost any paper, ceramic, or mica capacitor: Is it open, or is it leaky or shorted? Measuring the capacitance value is seldom necessary, since this information normally is stamped or color-coded on the capacitor itself. Capacitors of these types are not likely to shift in value, so the real question boils down to whether a capacitor is good or bad.

A capacitor tester is handy for determining whether a capacitor is open. Hook it up and turn the dial rapidly from one end to the other, past the nominal value of the capacitor. If the tuning eye on the tester opens at all, the capacitor is not open.

Make the same check with an ohmmeter for values larger than .01 $\mu$F by touching the ohmmeter across the capacitor and watching for the charging kick. (Smaller values give a charging kick, too, but it is too small to show on the meter.)

Next, and most important, check the capacitor for leakage. A dead short or high leakage can be caught with the ohmmeter. If the ohmmeter shows any deflection on the highest ohms range available, the capacitor is bad. For very critical applications, such as audio coupling capacitors in vacuum-tube amplifiers, you need a test instrument that reads very small leakage. Even a leakage of 100 M is enough to cause trouble in a coupling capacitor.

The fastest test for a possibly open capacitor is to bridge another one across it. If you suspect oscillation is caused by an open bypass capacitor, for example, bridge it. If the oscillation stops, you have found the trouble. If a coupling capacitor is suspected, you can test it in two ways: check for the presence of signal on both the input and output sides of the capacitor; or bridge another capacitor across it — if the signal now goes through, the original is open. If a capacitor opens, it has the same effect as taking the capacitor completely out of the circuit. So, replace it by bridging, and see if the trouble stops.

There has been a lot of development in capacitance meters in recent years. Newer devices check for opens and shorts, measure leakage, and display the capacitance value directly. These devices and other new types of test equipment were discussed in chapter 1.

## Electrolytic capacitors

Electrolytic capacitors, unlike paper capacitors, can change value by drying up. A dry electrolytic is not dry any more than a dry-cell battery is. If the electrolyte evaporates in either one, it stops working. A battery dies, and a capacitor opens completely.

With electrolytics, the best test is again, "How well do they work?" If the service notes specify that a power supply should have 275 V at the rectifier output and you find only 90 V to 100 V, the input capacitor is very likely open. Bridge it with a good one; if the voltage jumps up to normal, that was the trouble. If the ripple or hum level is given as 0.2 V p-p at the filter output and your scope shows 10 V to 15 V p-p there, bridge the output capacitor. If the ripple drops to within the proper limits, that capacitor was open.

For bridging purposes, the test capacitor does not need to be an exact duplicate of the original. It can be much larger or smaller in value, but it must have a working voltage able to stand whatever voltage is present in the circuit. It is not a good idea to bridge electrolytics in transistor circuits with the amplifier on. The charging-current surge of the test capacitor can cause a sharp transient spike in the circuits, which can puncture transistors. So, to bridge-test transistors sets, turn the set off, clip the test capacitor in place, and then turn the set on again. With the instant starting of transistor circuitry, this won't waste any time.

If you find one unit in a multiple-type electrolytic to be bad, replace the whole can. Whatever condition existed inside that can to make one unit go out will eventually cause failure of the rest because they're all parts of the same assembly. To avoid an almost sure callback, change the whole thing at once.

## Finding the value of an unknown capacitor without a capacitance meter

If you don't have a modern capacitance meter handy, there is still a fairly quick way to find the value of an unknown capacitor.

Note: This method isn't accurate unless you have a precise ac voltmeter or calibrated scope and an accurate test capacitor. The method is handy for finding the values of those odd mica capacitors that everyone has around and can't read the color code on.

Put the capacitor in series with a known capacitor, as shown in Fig. 6-4, and apply an ac signal voltage across the two. Measure the voltage across the known capacitor, then measure the voltage across the unknown capacitor. The voltage ratio will give you the ratio of the capacitances.

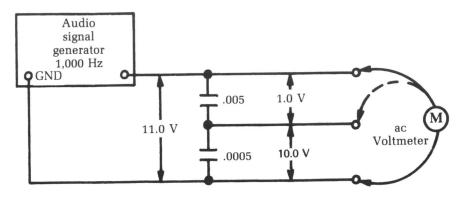

**Fig. 6-4**  *A capacitor's value can be approximated with a test hook-up like this.*

What you're doing is putting two reactances in series across an ac-voltage source. The result is a voltage divider that works the same way as two resistors in series across a dc source. Suppose, for instance, that you put 11 V at 1,000 Hz across a series combination consisting of a .005 $\mu$F capacitor and an unknown capacitor. Suppose then you measured 10 V across the unknown and 1 V across the known. The voltage ratio for the unknown and known capacitors is 10:1, and so is the ratio of reactances. The ratio of *capacitances* is just the opposite, or 1:10, because the reactance of a capacitor is inversely related to the capacitance: As capacitance increases, reactance decreases, and vice versa.

In the example given, the unknown capacitor has a value one-tenth as great as the known capacitor, or .0005 $\mu$F. Remember, a voltage ratio of 10:1 means a capacitance ratio of 1:10. If you prefer, you can work with reactance values, reading them from a table and then using the table to find the corresponding capacitance values.

### Testing capacitors with a scope

An oscilloscope is another handy tool for checking capacitor problems. Generally the standard low-capacitance probe is sufficient, although in some cases, a demodulator probe is useful.

To test the capacitor, adjust the scope for maximum sensitivity on vertical gain. Then touch the probe tip to the hot side of the capacitor to be tested. Interpretation of the scope trace depends on the capacitor's intended application in the circuit. In most circuits, capacitors are used for bypassing, filtering, or coupling.

A bypass capacitor is always grounded at one end. Unwanted ac components of a signal are shunted to ground. In this test, you want to determine that the ac components have indeed been eliminated from the signal.

A filter capacitor is really very similar to a bypass capacitor, although it is usually somewhat larger in value. A filter capacitor's function is to shunt unwanted signals, such as ac hum or RF signals, to ground. It is easy to determine if these components are present in the scope trace.

A coupling capacitor, or *blocking capacitor*, is placed in the circuit to pass an ac signal but to block any dc component in the signal. Check the scope trace to make sure the ac signal is centered around 0 and is not riding on a dc voltage.

# Testing variable capacitors

Variable capacitors are not used in modern electronics nearly as often as they once were, but you still might encounter them from time to time, especially in older radio tuners. In a typical variable capacitor, a set of movable plates, or rotors, are moved either directly via a tuning knob, or indirectly with a dial cord and a system of pulleys. These rotors are intermeshed with a second set of fixed-position plates, known as *stators*. The rotors move with respect to the stators. The air between the two sets of plates acts as the capacitor dielectric, so the relative positioning of the two sets of plates determines the effect capacitance of the component at each setting.

Test the resistance from any of the rotor plates to the metal frame of the variable capacitor. If the circuit is mounted on a metallic chasis, the component might or might not be insulated from the chassis. You should measure a dead short, or 0 $\Omega$.

Disconnect the wires connected to the stator terminals and measure the resistance from each stator plate to the rotor and/or the unit's metal frame. You should get an open circuit reading. The ohmmeter should indicate infinity, or a very, very high resistance. To be doubly sure, it is a good idea to make this test at each extreme of the tuning shaft's range; that is, once with the plates fully meshed, and once with them as opened up and far apart as possible. You also should get an infinity reading between individual stator sections.

In servicing equipment with variable capacitors of this type, you also must watch out for mechanical defects. It is usually very easy for one or more of the capacitor's plates to get bent. A bent plate can restrict the movement of the tuning control. It also can create a short circuit between the rotor section and the stator sections, causing noise or a partial or complete loss of signal. In some cases, turning the tuning control will have no apparent effect because of such a mechanical short circuit.

These tests are only for open-air types of variable capacitors. They are not appropriate for some other types of variable capacitors.

# Testing inductors

A coil, or *inductor*, is basically a pretty simple component. It is really nothing more than a length of wire wound in a coil shape around a core of something. (In some cases, there is an air core. That is, there is nothing at the center of the coil except empty air.) This type of component isn't used in modern electronics nearly as much as it used to be, but it is still far from uncommon. You won't run into as many coils as capacitors or resistors, but the odds are, you will run across some in the electronic equipment you are servicing.

The basic unit of inductance is the henry (H), or, more commonly, the millihenry (mH). One henry equals 1,000 millihenries. Unfortunately, standard electronic test equipment cannot measure inductance values directly (see the next section of this chapter). Fortunately, you can do some crude, but effective, good/bad tests with an ohmmeter to find most common faults with this type of component.

As with most component resistance readings, you probably will have to disconnect at least one end of the inductor from the circuit to avoid the effects of parallel resistances that can confuse the picture. However, this problem is generally less significant for coils than for other types of electronic components because the resistance of a typical inductor is so low that it will probably be the major factor in determining the total effective parallel resistance. Still, there are exceptions.

There are essentially three basic ways a coil can go bad: It can be open, there can be a short between windings, or there can be a short to the inductor coil.

## Open coils

Open coils aren't very common, but they can happen. An open coil is just a break in the coiled wire, or more likely, in its leads, which connect it to the rest of the circuit. If you measure the resistance across the terminals of a coil and get an infinity (or very high) reading, then you have an open coil. In some cases, it might be worthwhile to take just a moment or two to see if you can locate the break visually.

In some cases of a broken lead, you can resolder it without too much trouble. Otherwise the coil is defective and must be replaced. In most practical servicing situations, there is little point in spending much time in trying to locate the actual break, since the odds are that it won't be reparable.

## Shorted turns

It is trickier to determine whether or not there are one or more shorted turns in an inductor. Most practical coils consist of many closely wound turns of fine wire. If some of the insulation chips off, melts, or is otherwise removed, the conductive wires of adjacent turns will touch, producing a short circuit, as illustrated in Fig. 6-5. The coil will act like it has too few turns, and therefore, it will not function properly in the circuit.

If you know the nominal resistance of the coil in question, you can simply measure the resistance across the coil's terminals. If you get an abnormally low resistance reading, then a short within the coil is probable. Replace the defective inductor.

The nominal resistance of coils often is included in service data or specification sheets. If it is not, however, it is a lot harder to determine what the correct resistance for the coil should be. Normally, it will be a very low value, since, as far as dc current (and resistance) is concerned, a coil is nothing but a long length of ordinary wire. A dc signal couldn't care less that the wire is coiled, rather than straight. Any conductor has some finite resistance per foot. The longer the conductor, the lower the resistance. Shorted turns in a coil effectively shorten the conductor length.

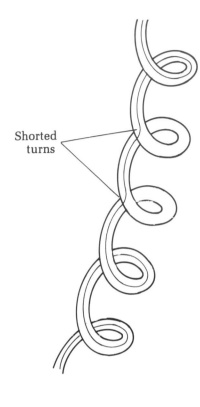

Shorted
turns

***Fig. 6-5*** *Shorted turns in a coil effectively make it function like a smaller inductance.*

If the circuit contains more than one coil, you might compare the resistance readings across each of them to see if they are reasonably similar. Don't make too many assumptions. There is no guarantee that all similar-looking coils in a circuit have the same value. One might have 100 turns, while another might have 225 turns, so their resistance readings logically will be quite different. You can, however, sometimes make some reasonable guesses. For example, let's say you have a circuit with four coils that look pretty much alike. When you measure the resistance across each coil, you get the following readings:

| Coil | Ohms |
|:----:|:----:|
| A | 51 |
| B | 48 |
| C | 11 |
| D | 55 |

With these readings, I would certainly be suspicious of coil C, especially if the schematic indicates that all four coils should have similar inductance values.

## Shorts in inductor coils

The third common type of defect in an inductor is a short to the core. This is mechanically similar to the shorted turn discussed earlier. One or more of the windings in the coil has lost some of its insulation for some reason, and since the turns of wire are pressed tightly against the coil, there is an electrical short circuit between the core and the exposed winding(s). Obviously, this is only a problem in an inductor with a conductive core, such as an iron core. You won't encounter this type of problem with an air-core coil, of course.

You can determine if there is a core short by measuring the resistance between either terminal of the inductor and the core itself. Ideally, this resistance should be infinite. In most practical cases, you will get a finite, but very high resistance reading. A typical value will be over 25 M$\Omega$. The exact winding to core resistance depends on the core material and configuration, as well as the quality of the insulation used in the construction of the inductor.

These same tests also apply to RF chokes, since this type of component is just a special-purpose coil. The term *choke* has more to do with the function than the actual construction of the component itself.

You can test a transformer in a similar way. A transformer is made up of two (or sometimes more) coils wound on a common core so their electromagnetic fields interact. Test the primary winding as if it was a simple coil. Then do the same with the secondary winding(s). Most transformers have just a single secondary winding, but some have more.

Test each secondary winding separately, but don't get confused by center taps. A *center tap* is an extra terminal to permit circuit connection to a midpoint on the coil. A center tap is clearly shown in the standard schematic symbol for a transformer with a center-tapped secondary, as illustrated in Fig. 6-6. Figure 6-7 shows a transformer with three secondary windings. Notice that one of these secondaries also has a center tap.

You also should test for electrical isolation between the primary and each secondary winding of a transformer. Measure the resistance between either end connection of the primary winding and either end connection of the secondary winding. (For a transformer with more than one secondary, repeat this test for each secondary winding.) You should get a very high (nominally infinite) resistance reading. If you get a resistance reading of less than

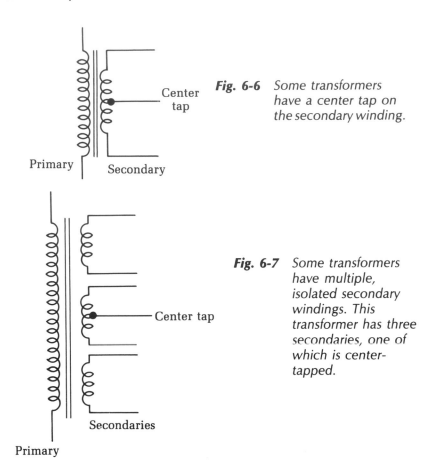

**Fig. 6-6**  *Some transformers have a center tap on the secondary winding.*

Center tap

Primary    Secondary

**Fig. 6-7**  *Some transformers have multiple, isolated secondary windings. This transformer has three secondaries, one of which is center-tapped.*

Center tap

Secondaries

Primary

20 MΩ or so, suspect trouble. There might be a partial or total short between the windings, which will interfere with the correct operation of the transformer and could easily result in some dangerous conditions. A very low reading here is a clear indication that the transformer should be replaced immediately.

Do not attempt to operate any electrical equipment with such a defective transformer. If you're lucky, there will only be major damage to the rest of the circuitry. However, there is also a very real possibility of dangerous, and potentially fatal electrical shocks and/or fire.

One important exception to this last test. A special type of transformer, known as an *autotransformer*, uses the same coil for both the primary and the secondary, as illustrated in the schematic symbol shown in Fig. 6-8. Obviously, in this case you inevitably will read a very low resistance between the "primary" and the "secondary," since they are just different points along the very same length of coiled wire.

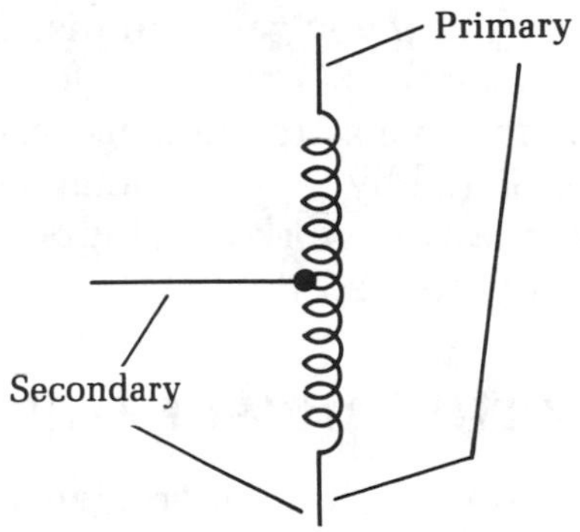

**Fig. 6-8**  *An autotransformer uses the same physical coil as both the primary winding and the secondary winding.*

Primary

Secondary

# Measuring inductance

The question is often asked, "How can you measure inductance?" The best answer is, "You can't." Inductance measurements, with common shop equipment, are a practical impossibility. You can figure it out by spending lots of time and doing a lot of mathematics, but the best advice we can offer is, "Don't."

In practical service work, there's seldom a need to read the inductance of a coil or transformer in henrys or millihenrys. You are usually interested in just one thing: continuity. This is a simple ohmmeter test. All service data give the dc resistances of coils, and as long as your resistance reading on an inductor is within 5 percent of the specified value, the inductor is probably all right.

Only two things that can happen to an inductor: It can open completely (which is fairly easy to find), or it can develop shorted turns. In power transformers, etc., shorted turns give a very definite indication: smoke. In other circuits, such as output transformers, shorted turns cause a drastic loss of output. You can locate a shorted turn by elimination tests and power measurements.

Flybacks are a special case; they act more like tuned circuits than ordinary transformers do. You can test them with a special instrument that connects the coil into a circuit and makes it oscillate. Read the Q of the coil on a meter. However, you can check flybacks for shorts by reading the cathode current of the horizontal-output tube and then disconnecting all loads, such as the yoke and damper circuits, and reading again. If the cathode current is far above normal, the flyback is internally shorted. When the loads are disconnected, the current should drop to about one-fourth its normal full-load value.

To get an inductance of any particular value, there's one easy way: buy it. You can wind coils all day, trying to get an 8.3 $\mu$H choke, but if you call your distributor, you can have a choke coil with exactly 8.3 $\mu$H in a few minutes. Coil and transformer makers have a tremendous selection of coils in all conceivable sizes, listed by their inductance, mounting style, etc. By far, the easiest way to work with inductors is to go buy an exact duplicate when you need it.

# Shorted power transformers

A common source of problems in electrical circuits is the short circuit. The trick is in determining just where it is. It is usually somewhere in the load circuit, but infrequently, the problem might be in the power transformer. You could waste hours going through every single component in the load circuit and tearing your hair out because everything checks out fine, when the problem is really in the power supply itself. Be aware of this possibility, although it is definitely the exception, rather than the rule.

A handy "quick-and-dirty" test is to disconnect everything from the secondary (or secondaries) of the power transformer. When you do so, the transformer should drive no external load at all. Next, plug in the device and wait 10 to 20 minutes. Feel the case. Is it abnormally warm? Some heat is to be expected, but if it is abnormally warm under these conditions, the power transformer is probably shorted. If the case is too hot to touch comfortably, there is definitely a problem with the power transformer.

You've narrowed down the problem to the power transformer, since nothing else is getting power, so no other component can be the source of the excessive heat. Nor can the transformer be overloaded, since it is running with no load at all, so the current drawn from its secondary winding(s) should be zero. The only possible cause for this heating is some shorted windings within the transformer itself. The transformer must be replaced or rewound.

You can use a dynamometer-type wattmeter for a more exact test. A good power transformer with no load should give a very small rating, only 5 W or less in a large power transformer. This small wattage is a result of the normal iron loss effects within the transformer, which is why some heat is inevitable. A higher wattage reading with no load is a clear indication of trouble.

Not all wattmeters will work properly for this test. You must use a dynamometer type, with both a voltage coil and a current coil. Such a wattmeter will have four terminals, instead of just two. If you don't have a suitable wattmeter handy, you can make a simple, crude, but reasonably effective wattmeter with an ac voltmeter (the ac volts section of your multimeter) and a 1 Ω power resistor in series with the ac circuit to be monitored.

## Checking the turns ratio of a transformer

A power transformer consists of two coils closely wound around a single core. Passing an ac voltage through one coil (the primary) will induce a proportionate ac voltage in the other coil (the secondary). Note that some transformers have multiple secondary windings with a single primary. The induced voltage across the secondary winding might be the same as the input voltage applied to the primary winding, but usually it will be different. The relationship between the input and output voltages is determined by the relative number of turns in each winding. There are three basic combinations:

| Type | Output voltage compared to input voltage | Secondary turns compared to primary |
|------|------------------------------------------|-------------------------------------|
| Step-down transformer | lower | fewer |
| Isolation transformer | same | same |
| Step-up transformer | higher | more |

The turns ratio is defined by the number of turns in the primary winding divided by the number of turns in the secondary winding. Since this ratio will correspond to the proportional voltages, you can also define the turns ratio in terms of the primary and secondary voltages:

$$\text{Turns ratio} = \frac{\text{Primary voltage}}{\text{Secondary voltage}}$$

Let's see how this works with a few quick examples. In all cases, assume an input (primary) voltage of 120 V.

First, let's say the output (secondary) voltage is 24 V. This is a step-down transformer with a turns ratio of:

$$\text{Turns ratio} = \frac{120}{24}$$
$$= 5 \text{ greater}$$

The turns ratio of a step-down transformer is always greater than 1.

For the second example, the output (secondary) voltage is also 120 V. This is an isolation transformer. The turns ratio in this case is:

$$\text{Turns ratio} = \frac{120}{120}$$
$$= 1$$

The turns ratio of an isolation transformer is always exactly 1.

Finally, let's look at a step-up transformer with an output (secondary) voltage of 500 V. This time the turns ratio works out to:

$$\text{Turns ratio} = \frac{120}{500}$$
$$= 0.24$$

The turns ratio of a step-up transformer is always less than 1.

In practical electronics work, when you run across a transformer that may or may not be bad, it is helpful to know either the transformer's rated primary and secondary voltages or the turns ratio. (In the United States, it is generally safe to assume that the primary voltage of a power transformer is probably 120 Vac.) If just a few turns are shorted in one of the transformer's windings, it might look fine using the resistance test described earlier in this chapter, but there could still be a significant difference in secondary voltage for the rated primary voltage because the turns ratio has been altered.

As an example, let's use the 120 V : 24 V step-down transformer from an earlier example. You already know the turns ratio of this transformer is 5. For every turn in the secondary winding, there are five turns in the primary winding. Assume there are 100 turns in the primary winding. This means there must be 20 turns in the secondary winding.

Now, let's say four turns in the secondary winding are shorted. This will remove less than an inch from the effective length of the conductor, so the difference in the resistance probably will be negligible. It may even be indetectable. However, the turns ratio of the transformer has been changed. Instead of 100/20 (5), the new turns ratio is:

$$\text{Turns ratio} = \frac{100}{16}$$
$$= 6.25$$

By rearranging the turns ratio equation, we can prove that the output (secondary) voltage of the transformer is equal to the input (primary) voltage divided by the turns rattio. That is:

$$\text{Secondary voltage} = \frac{\text{Primary voltage}}{\text{Turns ratio}}$$
$$= \frac{120}{6.25}$$
$$= 19.2 \text{ V}$$

The output (secondary) voltage of this transformer will be too low because of a few shorted turns in the secondary winding.

Fortunately, it is not too difficult to perform a direct voltage test on a transformer. First, remove the ordinary circuit load. Place a resistor across the secondary winding to serve as a simple known load. To determine the proper resistor value, you need to know the intended output (secondary) voltage and the approximate current drain of the normal load. You can often get a good estimate current value from the circuit's fuse rating.

Once you know these factors, at least approximately, just use Ohm's law to determine the value of the load resistor.

For example, let's say a 24 V step-down transformer normally drives a load that draws 50 mA (0.05 ampere). The required load resistance in this case is equal to:

$$R = \frac{24}{0.05}$$
$$= 480 \ \Omega$$

A standard 470 $\Omega$ resistor will probably be close enough.

Be careful. In most cases you cannot use a standard ½ W or ¼ W resistor. You will probably need a larger power resistor. To determine the required wattage rating, just multiply the voltage and the current:

$$P = E \times I$$
$$= 24 \times 0.05$$
$$= 1.2 \text{ W}$$

In this example, the load resistor must have a wattage rating greater than 1.2 watts.

The next step in the test procedure is to measure the actual input (primary) voltage. Sometimes ac power lines might run a little high or, more commonly, a little low. Of course, this difference will affect the secondary or output voltage. For example, if a transformer is rated for an output of 24 V with an input of 120 V (turns ratio = 5), but the actual input voltage is only 110 V, the actual secondary voltage will be reduced to:

$$\text{Secondary voltage} = \frac{\text{Primary voltage}}{\text{Turns ratio}}$$
$$= \frac{110}{5}$$
$$= 22 \text{ V}$$

Now, measure the voltage drop across the load resistor, as illustrated in Fig. 6-9. Do you get the expected secondary voltage? If you get a reading more than about 10 percent off from what it should be, the transformer is probably defective.

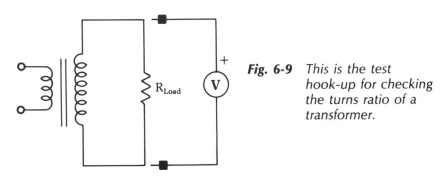

**Fig. 6-9**   *This is the test hook-up for checking the turns ratio of a transformer.*

## Rewinding transformers

Usually, when a transformer tests bad, you will be inclined to simply discard it and replace it with a new transformer of the same specifications. Unfortunately, it is sometimes difficult to find an exact replacement for a specific transformer. Occasionally, you will run into a problem with physical size. The available replacement transformer might have the same electrical specifications as the original unit, but if it has a larger body, it might not fit into the available space in the equipment being serviced.

Even if you do find a suitable replacement transformer, these devices are generally rather expensive. In some cases, it might make good economic sense to try to repair the defective original transformer by rewinding it, rather than simply replacing it. Another reason you might need to rewind a transformer is if you can't find one with quite the required turns ratio. By rewinding the secondary, you can redesign the transformer's specifications to suit your requirements. To aid you in such repairs, this section will discuss the procedure for rewinding a transformer.

Note that rewinding a transformer is a very picky, time-consuming job. It takes a lot of patience, good concentration, and a steady hand. Many electronics technicians simply aren't temperamentally suited for such work. For them it would always be worth the extra expense to seek out and purchase a replacement transformer of the required type. Before attempting such a repair to save on the cost of a new transformer, you should also consider what your time is worth.

We will only cover the procedure for rewinding the secondary winding of a transformer. It is possible to rewind a transformer's primary winding, but this procedure is almost always more trouble than it's worth.

Try to avoid attempting such repairs on transformers with multiple secondary windings. The design of the specific transformer in question also will help determine the suitability of such a repair attempt. In some transformers, the secondary winding is on top of the primary winding. Such transformers are usually rewindable. Other transformers, however, have the primary winding on top of the secondary winding. This arrangement is more common in step-up transformers than in step-down transformers. Repairing this type of device would be excessively difficult and time-consuming. It probably would not be worthwhile.

When attempting to rewind a transformer, it is vital to work slowly, carefully, and gently. Never use force.

If you intend to change the turns ratio, you must use a transformer with the correct core size. The more power passing through the transformer, the larger the core must be. A chart for roughly determining the cross-sectional area of a transformer core is given in Table 6-1. Of course, if you are reducing the secondary voltage, there should be no problem. These cores sizes are the minimum for each wattage level.

Electrically, it doesn't matter if the transformer's core is too big, although the device will be physically bulkier than it needs to be. The key point is that you cannot reasonably expect to make too large a change in the original transformer's output power. For example, if the transformer was originally rated for 24.6 V at 1.5 amps, you should be able to rewind it for 30 V at 1 amp, or 18 V at 1.75 amps, but not 35 V at 3 amps. The core would be too small, and the transformer would burn itself out and possibly damage other components in the circuit. It also might be a fire hazard.

**Table 6-1   The cross-sectional area of a transformer's core limits its power-handling capability.**

| Cross-sectional area in square inches | Maximum power  in watts |
|---|---|
| 1.00 | 45 |
| 1.25 | 50 |
| 1.50 | 65 |
| 1.75 | 75 |
| 2.00 | 120 |
| 2.25 | 150 |
| 2.75 | 230 |
| 3.00 | 275 |
| 3.25 | 330 |
| 3.75 | 440 |
| 4.00 | 520 |

To begin the rewinding process, you must first open up the transformer. Remove all screws or anything else holding the transformer together. The body of the transformer (the core) is made up of multiple sheets of laminated metal. During manufacture, these laminations were soaked in a special enamel and then baked, so they are tightly sealed to one another. The enamel protects the transformer from environmental contamination and helps prevent annoying transformer buzz. Unfortunately, it also makes it difficult to remove the laminations.

You must individually break the coat of enamel holding each lamination in place. Be very careful when doing so. If you use too much force, your tool could slip and damage the transformer's wires. Too much damage to the wrong wires could render the transformer unrepairable.

You are likely to damage a few of the laminations themselves, especially the first few, which are almost always the hardest to remove. They have the most enamel, since they are on the outside. If you ruin a few laminations, don't worry too much about it. Because you will be reassembling the transformer by hand, all the original laminations won't fit back into place anyway. Just try to work as carefully as possible and not damage too many of the laminations.

The best way to remove the first few laminations is to use a small screwdriver. Attempt to work the blade between the topmost lamination and the next one. A few gentle taps with a hammer can help push the screwdriver's blade between the laminations. Do not use hard blows. Be aware that this process is likely to damage the blade of the screwdriver, so do not use an expensive, precision tool. A small, cheap screwdriver is not hard to find and will do the job. The blade should be as narrow as possible.

Work the blade gently back and forth between the laminations until you manage to break them apart. This procedure is difficult until you get the hang of it.

After you have removed a few laminations, you will almost certainly come across some that are I shaped. An I-shaped lamination will do a better job than the screwdriver, so make the switch as soon as you can.

After removing the laminations, you can unwind the secondary of the transformer. Do so slowly and carefully, at a time and place where you are unlikely to be disturbed in the middle of the job. Be careful not to break or tangle the wire, especially if you intend to reuse it in the new winding. Keep track of the number of turns in the secondary winding. To rewind the transformer for the original specifications, you will need to use exactly the same number of turns. If you intend to change the turns ratio, you will still need to know how many turns there were so you can perform the necessary conversion.

If you are repairing a transformer with shorted windings, you should not attempt to reuse the original wire in the new secondary winding. Use enameled wire of the exact gauge as the original. You also will need to use new wire if the new secondary winding calls for more turns than were used in the original transformer.

To determine the number of turns required for a different secondary voltage, determine the original number of turns per volt. Just divide the total number of turns in the original secondary winding by the original output voltage:

$$TV = \frac{T_o}{E_o}$$

where $TV$ is the number of turns per volt, $T_o$ is the total number of turns in the original winding, and $E$ is the original secondary voltage. As an example, let's say you have a 24.6 V transformer with 125 turns in its original secondary winding. The turns per volt for this transformer is:

$$TV = \frac{125}{24.6}$$
$$= 5.0813$$

Now, if you want to rewind this transformer for a new secondary voltage of 18 V, just multiply the desired voltage by the turns-per-volt value:

$$NT = En \times TV$$
$$= 18 \times 5.0813$$
$$= 91.4634$$

where NT is the number of turns required for the new secondary winding, En is the desired output voltage, and TV, of course, is the turns-per-volt value derived previously.

Don't worry too much about the fractional turns. In this example, 91.5 turns would certainly be close enough. Actually the difference between 90 or 91 turns and 92 or 93 turns will be negligible. Make the new winding as close as possible to the calculated number of turns, but don't be obsessive about it.

The current-handling capability depends on the wire's cross-sectional area. To use the transformer for more than its originally rated current-handling capability, you might need to use a heavier gauge wire. I don't really recommend doing so.

In most cases, you should use new enameled wire for the new secondary winding. The original wire might have physical strain and kinks as a result of being tightly wound so long, then being unwound and rewound. There also might be chips in its insulating enamel, which could eventually result in shorted turns. If you were very careful in unwinding the original coil, however, you might be able to reuse the original wire, if you wish. Do not reuse the original wire if you are repairing a defective transformer, especially one with shorted turns. You'll just be rebuilding a defective transformer.

Wrap the wire around the core as closely as you can. Do not leave any space between turns so you get the maximum number of turns in the minimum amount of space. We guarantee you won't be able to get the windings as tight as they were in the original manufactured unit, so your rebuilt transformer is likely to be a little bulkier. This is okay.

When you finish one layer of turns, put some kind of insulation over it before you begin the next, overlapping layer. You can use wax paper, electrician's tape, or any other thin, insulating material, provided that it is capable of withstanding the maximum voltage of the winding. Apply this insulating material as tightly and as wrinkle- and bubble-free as possible to prevent excessive bulk.

Once you have completely wound the new secondary, you must reassemble the transformer's laminations. Some will be E shaped; others will be I shaped. They fit together as illustrated in Fig. 6-10. Because you are working by hand, and the transformer was originally machine-assembled, it is almost a sure bet that you'll end up with a few laminations left over—typically about two to four of each type (E and I) if you are good at such delicate work. If you can't fit these extra laminations back into the transformer's core, just discard them and don't worry about it. These omitted laminations will lower the power rating of the transformer slightly, but the difference is not likely to be significant or even noticeable. Of course, if you end up with a dozen or more laminations left that won't fit back into the core, you better derate the power-handling capability of the rebuilt transformer somewhat.

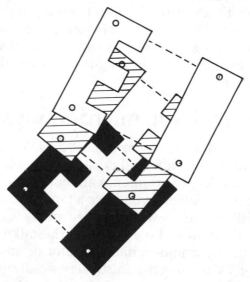

**Fig. 6-10**  *The laminations of a transformer's core fit together like a simple puzzle.*

Finally, replace all the original screws and other hold-down devices that you removed from the original transformer.

# Testing integrated circuits, modules, and PC units

Printed-circuit units are appearing in radios and TV sets in large numbers. These devices range from a simple RC integrator used in vertical sync circuits to the equivalent of a whole amplifier circuit, each in one sealed package. Admittedly, these units are impossible to check in detail because you can't get into them to test individual parts. However, there is at least one reliable check you can do: an output check.

In the integrated circuit, check to see that the proper composite sync signal is present at the input. If it is not found at the output, the unit is probably bad. This method can be used on any module. Three things must be carefully checked before any printed-circuit unit is condemned: the supply voltages and currents; the input signal; and the output signal.

For instance, if the modular circuit is an audio amplifier, it will need a certain amount of dc voltage supply and draw a certain amount of current. With 0.5 V of audio signal on the input, it has a normal output of 5 V. If the unit meets these specifications, look elsewhere for the trouble. Don't replace units at random. Make definite tests and be positive before you replace any units.

The scope and signal generator can tell you if a stage is definitely bad by checking input vs. output. If it is bad, don't overlook the supply voltage.

# A quick check for microphones

There's a good, quick check for almost all microphones, especially the common dynamic and crystal types: Make them talk, rather than listen. Any microphone can reproduce sound, as well as pick it up. For instance, if you have a tape recorder that won't record, the first question is whether the trouble is in the mike or the amplifier. Feed an audio signal into the mike and listen.

You can use an audio-signal generator or any audio signal from a radio or TV set. It takes only a very small signal to make a mike talk. A dynamic microphone is nothing but a specially built dynamic speaker. Crystal mikes won't talk as loudly as dynamics, but even the variable-reluctance types used in communications work will talk.

Incidentally, this is a good quality check for microphones if the complaint is distortion in the sound output of a PA system or transmitter. By feeding a music or voice signal into the mike and listening to it, you can detect dragging voice coils, buzzes, etc. — defects that would distort the sound pickup. Also, if you happen to have a replacement cartridge for the type of mike you are testing, you can easily make A-B comparison tests of the sound quality of each.

This test can be reversed, too. If you have a complaint of possible mike trouble, hook up a small dynamic speaker to the mike input and talk into it. If the mike input is high impedance, use an output transformer to bring the low voice-coil impedance up enough to work. Actual high-impedance mike transformers will be up around 50,000 Ω, but you can use almost any output transformer. One of the old 25,000 Ω transformers is good, but the test will work with even a 10,000 Ω type.

If you're checking for possible mike distortion, hold the mike at least 10 to 12 inches away as you talk. You'll be surprised at the quality of the sound. Transistor radio speakers make good test mikes because of their tiny size.

# Checking phono cartridges with the scope

When you have low gain in record-playing systems, one of the first things to determine is whether the trouble is in the cartridge or amplifier. The scope is a quick check for the cartridge. With its high-gain vertical amplifier, you can use it as a sensitive ac voltmeter.

Put a single-tone test record on the turntable, disconnect the cartridge leads at the amplifier (although this isn't really necessary if they're soldered in), and hook the vertical input of the scope to the hot wire, as shown in Fig. 6-11. Set the vertical gain control of the oscilloscope to give about 1 inch deflection for a 1 V p-p input. Put the stylus on a band of continuous tone — say 400 Hz. For the average crystal cartridge, the output will be from 1 to 3 V.

**Fig. 6-11** *An oscilloscope can be used to check a phono cartridge.*

If you want to make a frequency run on the cartridge, you can do it even with a narrow-band scope. Most of these scopes will go up to at least 50 kHz without trouble. You'll need a test record with a band of all frequencies on it, starting at 30 Hz and going to

20 to 30 kHz, at the same output level. Several test records of this type are available. You also can use this test on the whole amplifier.

One valuable application of a scope test is checking stereo cartridges for equality of output in the two channels. Use a test record with a monaural band at about 400 Hz or a stereo band with equal outputs in the two channels. Several test records have this band for checking speaker phasing, channel balance, etc. Just read the output from each side of the cartridge; the two should be the same.

Cartridge tests also can be made with an ac VTVM. The readings will be the same as with the scope: 1 to 3 V p-p. Even an AC-VOLTS scale on the VOM will do, although the meter must have a sensitivity of at least 10,000 Ω per volt on ac. The low input impedance of the VOM reduces the readings. They average from 0.3 to 0.4 V, where the scope reads 1 to 3 V. Crystal and ceramic phono cartridges should work into a load impedance of 3 to 4 M.

# Impedance checker for speakers

*Impedance*, or ac resistance, is generally difficult to measure. It is nowhere near as straightforward as dc resistance. For dc resistance, 100 Ω is 100 Ω, and that's that. It doesn't matter what signal is flowing through the resistive component. Impedance, on the other hand, is frequency-dependent. A component that has a 100 Ω impedance for a signal of 200 Hz might have an impedance of 38 Ω when the signal frequency is changed to 500 HZ.

For speakers, the situation is further complicated because the intended signal is a complex combination of multiple-frequency components. To make some sort of comparison possible, a standard test signal of 1 kHz (1,000 ohms) is assumed in defining the impedance of a speaker. The test signal is a pure sine wave, to avoid confusion from harmonic-frequency components.

The circuit shown in Fig. 6-12 is designed to test speaker impedance by comparing the speaker being tested with a known 8 Ω speaker. A suitable parts list for this project is given in Table 6-2.

IC1 and its associated components (R1 through R5 and C1 through C3) form a simple sine wave oscillator. The component values suggested in the parts list will give a signal frequency very close to 1 kHz.

Fortunately for this type of testing, high precision in the signal frequency isn't required. This simple circuit's frequency will be close enough for our purposes. If you prefer, you can replace

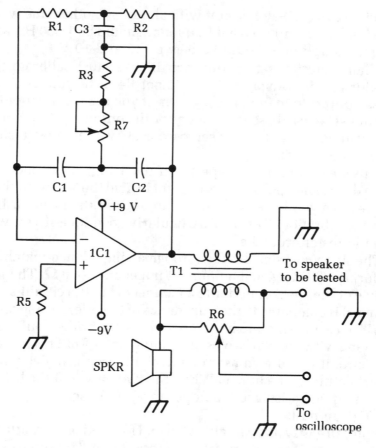

**Fig. 6-12**  *This circuit can be used to test the impedance of loudspeakers.*

**Table 6-2   The parts list for the speaker
impedance checker circuit of Fig. 6-12.**

| | |
|---|---|
| IC1 | 741 op amp |
| C1, C2 | 0.01 uF capacitor |
| C3 | 0.022 uF capacitor |
| R1, R2, R5 | 15K 1/2 Watt 10% resistor |
| R3 | 3.3K 1/2 Watt 10% resistor |
| R4 | 10K trimpot |
| R6 | 1K potentiometer |
| T1 | audio impedance matching transformer |
| | primary = 1K |
| | secondary = 8 ohms |
| S1 | small 8 ohm speaker |

this sine wave oscillator circuit with almost any other sine wave oscillator. The nominal signal frequency should be 1,000 Hz, and the signal amplitude should be between 3 and 10 V.

When you first set up this circuit, you must calibrate the oscillator circuit. Simply adjust trimpot R4 until you hear the clearest, purest tone from the speaker. If you are a perfectionist, you can use an oscilloscope to monitor the output of IC1. Again, adjust trimpot R4 until the oscilloscope shows the most distortion-free sine wave.

This is a set-and-forget type of control. It might be a good idea to use a drop of nail polish, glue, or paint to hold the trimpot's shaft in place once you have calibrated it to prevent the need for frequent recalibration. This is a particularly good idea if you will move the tester around a lot.

The sine wave signal is fed through an impedance-matching transformer (T1) to give it a nominal impedance of 8 $\Omega$. The impedance of the speaker under test (connected to the circuit's test terminals) is compared to the impedance of the reference speaker (8 $\Omega$). The two speakers to be compared are in a bridge configuration, along with the two halves of potentiometer R6. This portion of the circuit is redrawn as an equivalent circuit to give you a clearer picture of the bridge. When the two halves of the bridge have an equal resistance (or impedance), the oscilloscope will show a deep null.

You can easily make a calibrated dial for this tester by attaching known small-valued resistors across the test terminals and marking the dial for the null point. For example, you might start out with an 8 $\Omega$ resistor. While watching the oscilloscope, slowly adjust potentiometer R1 for the deepest null. Mark the potentiometer's position "8" on the dial. Repeat this procedure for resistances to match each of the standard speaker impedances: 3.2 $\Omega$, 4 $\Omega$, 8 $\Omega$, 16 $\Omega$, 32 $\Omega$, 40 $\Omega$, and 100 $\Omega$.

Now, when you connect an unknown speaker across the circuit's test terminals, you can adjust potentiometer R6 until you get the deepest possible null as indicated by the oscilloscope. Read off the approximate value from your calibrated dial. Sometimes you will get a speaker that falls between two standard calibration points. That's o.k. Just interpolate the value, or round it off to the nearest standard impedance value.

If you do not have an oscilloscope handy, you can substitute a pair of headphones and listen carefully for the maximum audible null. This method is a little less precise and requires more concentration. You also will have to make an effort not to be distracted by the tones being produced directly by the two speakers.

# Testing diodes

You can test a semiconductor diode with an ohmmeter, but you must be careful. Some semiconductor diodes are designed to handle only very small voltages. Too high a test voltage could damage or destroy the diode you are attempting to test.

To be safe, use an ohmmeter that uses a battery voltage of no more than 1.5 V. Some multimeters use up to a 9 V battery to power the ohmmeter section. It is a good idea to test unknown diodes on the highest range of your ohmmeter because, at these higher ranges, the ohmmeter has greater internal resistance, limits the current more, and puts less voltage across the diode connected to the test leads. We've found that, for most common semiconductor diodes, the upper resistance ranges give the easiest to read results anyway.

Measure the resistance across the diode from anode to cathode. Connect the ohmmeter's positive lead to the anode, and the negative lead to the cathode. This arrangement forward-biases the diode. You should get a very low resistance reading. The exact value will depend on the specific type of diode being tested. Some diodes have a forward-bias resistance of about 1 to 10 $\Omega$. Others might have forward-bias resistances of just a few tenths of an ohm. A few diodes might have higher forward-bias resistance.

Now, reverse the test leads. Connect the positive lead to the diode's cathode, and the negative lead to the anode. Now you should get a very high resistance reading because the diode is reverse-biased. On some diodes, the reverse bias might be just a couple hundred ohms, while others will exhibit resistances well into the megohm (millions of ohms) range. Most semiconductor diodes will have a reverse-bias range of at least several kilohms.

It is usually relatively easy to tell the anode from the cathode on a semiconductor diode. Often a stripe around one end of the component's body will indicate the anode, as illustrated in Fig. 6-13. Some diodes have a tapered body at the anode end, as shown in Fig. 6-14. Even if there is no visual indication at all, however, this test procedure will indicate which is which, assuming the diode is good.

For quick, general-purpose go/no go diode testing, you just want to make sure the diode exhibits a low resistance in one di-

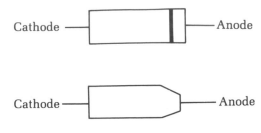

Cathode ———— Anode

**Fig. 6-13**  *On some diodes, the anode is indicated by a stripe.*

Cathode ———— Anode

**Fig. 6-14**  *On some diodes, the body is tapered on one end to indicate the location of the anode.*

rection (when it is forward biased) and a significantly higher re-sistance when the applied test voltage has its polarity reversed (the diode is reverse biased). If you get a low resistance in both di-rections, the diode is shorted and should be replaced. If you get a high resistance in both directions, the diode is open and should be replaced. For the most part, a semiconductor diode is either good or bad. When it goes bad, it is almost always shorted or open, so this test is sufficient.

In some cases, you might be concerned with the exact specifi-cations of the diode. The most important factor is the *front-to-back ratio* of the diode, a ratio of the forward-biased resistance to the reverse-biased resistance:

$$FBR = \frac{R_f}{R_r}$$

where *FBR* is the front-to-back ratio, $R_f$ is the resistance when the diode is forward-biased, and $R_r$ is the resistance when the diode is reverse-biased.

For example, if the forward-bias resistance is 75 $\Omega$, and the reverse-bias resistance is 12,500 $\Omega$ (12.5K), the front-to-back ratio is equal to:

$$FBR = \frac{12500}{7.5}$$
$$= 156.667$$

Is this reading good or bad? That depends entirely on the in-tended specifications for the diode in question. You must com-pare your test results with the manufacturer's specification sheet for that type number, or you can make direct comparisons with a diode of the same type that is known to be good. This example is probably a good diode. Usually if there are problems, they will re-sult in lower front-to-back ratios.

# Replacing transistors

It is often difficult to find an exact brand-name replacement for a bad transistor, especially one from an older, discontinued set. Foreign-manufactured equipment tends to include a lot of unusually numbered parts.

Many component manufacturers offer a line of general replacement devices, and all good technicians should have as many cross-reference guides as they can find. Unfortunately, it is not a good idea to place too much faith in any cross-reference guide. The recommended replacements are always just close approximations, not exact duplicates. A guide might list transistor A as a replacement for transistor B, and it will probably work great in 99 percent of the circuits using transistor B. But you may have the exceptional circuit. Also, don't assume that the replacements can work in either direction. That is, just because A is listed as a replacement for B, don't assume you could use B as a replacement for A.

The best cross-reference guides are ones that include data and specifications for each device. Generally, you will be most concerned with the voltage and current ratings and the cutoff frequency. The other specs are also important, of course, but in most applications they offer more room for error. When in doubt, or if you run into problems, use a replacement transistor with slightly higher ratings than the unit you are replacing.

# Derating components

A component circuit designer should take the power-handling capabilities of each component of the circuit into account. If the original component is rated for 250 V, don't replace it with one rated for only 100 V. Even if the circuit works correctly with the substitution, the underrated replacement component will be subject to premature failure. You'll just need to replace it again soon.

In some cases, the component originally installed by the manufacturer might be significantly overrated because of parts availability or some other reason. For example, we have seen ceramic disc capacitors rated for working voltages of 500 V in pocket radios that operate off of a standard 9 V battery. It is highly unlikely that the component will ever see any voltage coming even remotely close to 500 V in this circuit. In this case, you can safely substitute a capacitor with a lower voltage rating.

Usually things aren't quite so obvious. The general rule of thumb is to never substitute a component with a lower voltage or power rating than the original component unless you are absolutely sure that the manufacturer overrated the component for an electrically irrelevant reason (such as parts availability or cost). If there is any doubt, don't make the substitution. You might just be asking for trouble in the future.

On the other hand, you usually can overrate the voltage or power ratings of your replacement components. For example, you can replace a 2.2K, ¼ W resistor with a 2.2K, ½ W resistor. All you will have to worry about is whether or not the new, heftier component will physically fit.

In some cases, a manufacturer might have cut corners or the circuit designer might have erred, and the original component might not be adequately rated, or just barely so. If you service a number of units of a given piece of electronic equipment and repeatedly come up against a specific capacitor being burnt out, the odds are that its working voltage rating just isn't good enough for the operation of the equipment. In this case, you definitely don't want to use an exact replacement. Substitute a capacitor with the same capacitance value, but with a somewhat higher working voltage rating. This replacement will often give the equipment much greater reliability. We are constantly shocked by how often manufacturers use inadequately rated components, even in high-grade equipment.

As a rough rule of thumb, never expose any component to more than 75 to 80 percent of its maximum voltage or power rating during the normal operation of the circuit. For example, never expect a capacitor rated for 250 V to handle more than about 188 to 200 V, and a 0.25 W resistor shouldn't have to carry more than about 0.2 W in the normal operation of the circuit. This rule will leave some headroom for the component to handle any unexpected transients or overvoltage conditions without unnecessary damage.

# 7
# TV tests

LET'S BEGIN BY DISCUSSING DC CURRENT MEASUREMENTS IN THE SWEEP and high-voltage circuits. In all tube-type TV sets, both color and black-and-white, all the power used by the horizontal-sweep and high-voltage circuits is supplied *through* the horizontal output tube. You can tell a lot about the condition of such circuits by reading this current and comparing it to the normal value given on the schematic. Each likely trouble condition is indicated by its effect on the current. A leakage, short, or low-drive condition will make the current go up; an open circuit will make it go down. Disconnect parts to check the effect on the current to get the necessary "split" in the circuit to find a starting point for diagnosis. A typical circuit is shown in Fig. 7-1.

To measure the current of the horizontal output tube, put the tube on a special test adapter that breaks the cathode circuit. Then, connect a dc milliammeter in series with the cathode and read the total tube current. If you want to, deduct the 12 to 15 mA of screen-grid current to get the exact plate current. However, this step isn't usually necessary; use the total current.

Figure 7-2 shows how to build the test adapter. Clean out the base from a dead octal tube having all eight pins. Mount an octal socket on top. Run all pins except the cathode straight through — 1 to 1, 2 to 2, and so on. If you use solid No. 20 wire, it will hold the socket firmly on the base. If it becomes loose, you can cement the socket in place. Bring leads for measuring the cathode current out the side, as shown — one lead from the socket terminal, the other from the base pin. Use test-lead wire, which is very flexible and well insulated, for these. Put pin tips on the ends to match the jacks on your VOM.

Check the tube manual to make sure which pin is the cathode of the tube in the set you are testing. For the older tubes in black and white TV, such as 6BQ6, 6CU6, and 6DQ6, pin 8 is used.

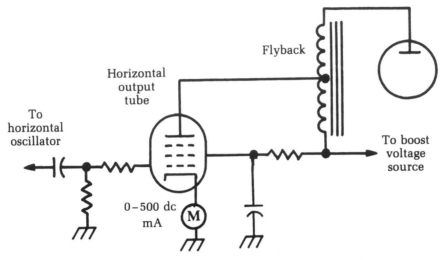

**Fig. 7-1**   *A typical TV high-voltage sweep circuit.*

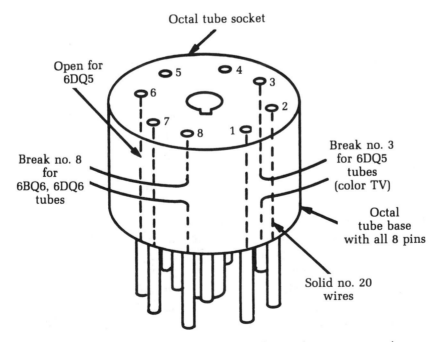

**Fig. 7-2**   *A simple home-brew test adaptor for vaccuum tubes.*

In the 6DQ5 tubes used in a great many color TV sets, pins 3 and 6 are the cathode. In the 9- and 12-pin types such as 6JE6, that are used in color sets built after about 1965, pin 3 is the cathode. (You'll probably have to buy special test adapters for these tubes because you can't use the tube base; it's made of glass. However, test adapters are available, and can be converted.)

In many sets, pins 3 and 6 are tied together at the socket; in most cases, pin 3 seems to be the favorite for making the ground connection. The service manuals specify opening this ground lead to hook up the milliammeter, but to do so, you must take the chassis out of the cabinet. With a suitable adapter, you can take the same reading from the top and save a lot of time.

Normal current will be specified. After a little practice, you'll learn the average current drain for the popular tube types. For example, 6BQ6 tubes should draw about 95 mA maximum; 6DW6 tubes can carry up to 120 to 130 mA safely; and 6DQ5s in color sets draw up to 185 to 200 mA. If you find a new tube type, check the tube manual to determine its safe cathode current, and to learn its pin connections.

## Measuring very high voltages

In black-and-white TV, it's uncommon to check the high voltage. Use the CRT screen as an indicator. If it's bright enough, assume that the high voltage is okay; it usually is. In color TV, however, the high voltage must be held within a fairly tight tolerance to avoid purity troubles, etc., so you must measure it. To do so, you need a dc voltmeter with a scale of at least 30,000 V. The easiest way to get such a range is to use an external multiplier probe with your regular dc voltmeter.

Multiplier probes are made with high-voltage insulating housings to keep the user's hands as far from the hot stuff as possible. The flanges are placed on the base to make the insulation path long. Inside the probe, a special high-voltage resistor — made by depositing a carbon film on a glass cylinder and then coating the whole thing with a high-voltage insulating plastic — is held between spring mountings that keep the contacts tight. The probe housing is sealed to keep out any moisture which might cause flashovers along the multiple resistor when the probe is in use. A VOM with a high-voltage probe is shown in Fig. 7-3.

If you have a 20,000 $\Omega$ per-volt VOM, you can set it to the 0 to 300 Vdc scale (to make the math simpler). What you need is a meter with a total resistance of 30,000 times 20,000, or $(3 \times 10^4)$

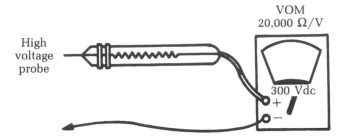

**Fig. 7-3**  *In dealing with very high voltages, it is necessary to use a high-voltage probe with the VOM.*

$(2 \times 10^4) = 6 \times 10^8 \, \Omega$, or 600 M. A 20,000 $\Omega$/V meter already has 300 $\times$ 20,000 or 6 M, on its 0 to 300 V scale, so subtract this number from the total. This leaves a probe resistance of 600 M — 6 M, or 594 M. With a 594 M probe connected, the 0 to 300 V scale of the meter is multiplied by 100 and reads 30,000 V full-scale.

You can use a multiplier probe with any meter, provided you use the correct multiplier resistance. Find the resistance of the VOM on the scale you want to multiply (full-scale voltage reading multiplied by the $\Omega$/V rating), then figure the total resistance you'll need to make the meter-probe combination read as high as you want. Subtract the meter resistance from this total. The remainder is the resistance you need in the multiplier probe. Probes are available in a variety of resistance values at radio distributor's stores.

For a VTVM or TVM, you have to use a slightly different method to determine the multiplier resistance. When you change ranges on a VOM, you change the meter resistance (0 to 3 V scale, 60,000 $\Omega$; 0 to 300 V scale, 6 M; etc.). On VTVMs and TVMs, the total input resistance stays the same for all scales. This resistance could be 11 M, 16 M, or whatever the instrument designer decided to use.

To figure the probe resistance, you use the same principle as before, but you get a different result. Say you have a 0 to 500 V scale and you want to make it read 0 to 50,000 V. This calls for multiplication — multiply by 100. First find the input impedance of the VTVM. Say that this is 11 M, a common value. Of this resistance, 1 M is in the regular DC-VOLTS probe, so the resistance of the internal voltage divider is 10 M. In most VTVMs, the regular volts probe is disconnected and the high-voltage multiplier probe plugged in its place, so the meter has a resistance of 10 M. To multiply readings by 100, multiply the input resistance by that figure, obtaining 1,000 M as the required meter-plus-probe resistance. Subtracting the 10 M you already have, the multiplier resistor in the probe will be 990 M.

Here's the difference between the VOM and the VTVM: since the input resistance on a VTVM never changes, multiply all the dc voltage ranges by the same figure when you hook up the high-voltage probe. Even the 0 to 3 V range becomes 0 to 300 V, etc.

If the regular probe resistor stays in the circuit when the high-voltage probe is attached, consider that fact when figuring out the high-voltage resistor value. In the case just figured, if meter resistance was 11 M, the multiplier would be 1,100 M — 11 M, or 1,089 M. One popular meter uses a 22 M input resistance with a 7 M resistor in DC-VOLTS probe. By removing this probe, you have 15 M left; so a 100:1 multiplier would require 1,500 M — 15 M, or 1,485 M. You'll find the exact input resistance of your own meter in the instruction book.

# Measuring focus voltages in color TV

In color TV, you often need to check the focus voltage. The best way is to take this reading right at the base of the color TV CRT to make sure that this important voltage is getting to where it is used. Also, check the action of the focus control to see if it gives the proper amount of variation. This variation is usually 4,000 to 5,500 Vdc. This voltage comes from the flyback, through a special rectifier—sometimes a small tube, sometimes a special high-voltage silicon rectifier.

In the original 21-inch color tubes and in many later ones, the focus electrode is pin 9 (check the schematic to make sure). You usually can tell just by looking at the set—many CRT sockets leave a blank space on either side of pin 9 to provide more insulation for the high focus voltage. The focus voltage is measured with a high-voltage probe, as illustrated in Fig. 7-4.

There's an easy way to make a focus voltage measurement. You need to make contact on the base pin of the CRT itself. This is hard to do if the CRT socket is pushed tightly on (as it should be). In some sets, you can pull the socket back just a little and get at pin 9 with a thin test prod. However, some of the other pins might not make good contact if you do. The answer is to make up a gadget that will let you get at this pin without disturbing any of the others.

One such device is a small clip made out of spring wire, as in Fig. 7-5A. Slip this clip over pin 9, and then push the socket back

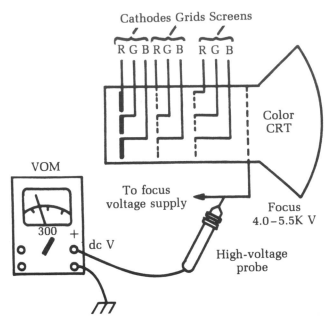

**Fig. 7-4**  *A high-voltage probe can be used to measure the focus voltage of a picture tube.*

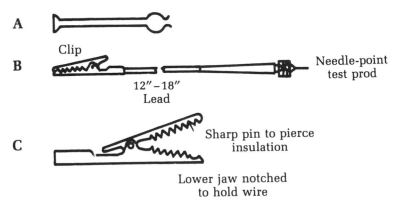

**Fig. 7-5**  *Different test clips are useful for different purposes: (A) Home-made wire probe for single-pin access. (B) Needle-point test-prod. (C) Clip with insulation-piercing pin.*

tightly. Touch the meter prod to the exposed end. (Just make sure that the meter prod is the *only* thing that touches; 5,000 V can bite!)

Another method is to use a needle-point test prod, as in Fig. 7-5B. Fit this prod to a short piece of test-lead wire with an alligator clip on the other end. Now lay the regulator high-voltage probe on top of the cabinet or on the bench, with the clip hooked to the end. You can use the insulation-piercing clip, too, as in Fig. 7-5C. The use of test clips like these leaves your hands free to adjust controls, etc., and keeps them as far as possible from that "hot stuff." Just be sure that the exposed ends of probes, etc., are far enough away from grounded parts so that there won't be a flashover when you turn the set on.

To check focus, turn the set on with the meter attached and move the focus control and see if it gives enough range of adjustment — 4,000 to 5,500 V for the average color set. Check the raster for normal, sharp focus of the scanning lines. In most sets, this occurs between 4,600 and 5,000 V; but in a few cases, you'll find voltages higher or lower than this amount. As long as you can get a good, sharply focused raster, it's okay.

If the focus voltage should be lower than about 10 percent, go to a higher scale and read the high voltage. The trouble could be caused by something in the horizontal output tube, flyback, damper, yoke, etc. If it is, then both focus and high voltage will be low by the same percentage. If focus voltage is down 40 percent and high voltage only 10 percent, look for trouble in the focus-rectifier circuits, etc. If the percentages are reversed — focus voltage only 10 percent low and high voltage 40 percent low — check the high-voltage rectifier, voltage regulator, and the circuits that affect only the high voltage output.

# Checking color picture tubes with a VOM

Your VOM can be a pretty good picture-tube tester, especially with color CRTs. One of the best ways to measure the quality of any picture tube is to read its beam current, which is the cathode current. It runs 300 to 400 $\mu$A maximum in most black-and-white tubes, and about the same for all three guns (total) in color picture tubes.

Many color TV sets have provisions for unplugging each cathode lead of the picture tube, as shown in Fig. 7-6, so the technician can adjust the beam currents of the three guns, if there happens to be a phosphor imbalance (this isn't as common as it once was). The three cathode leads are usually provided with push-on connectors that fit onto a terminal board on the back of the chassis. In some sets, you'll find the leads soldered to a terminal board.

As you can see, the red cathode lead goes through a small resistor, and the other two go to adjustable resistors. These are the

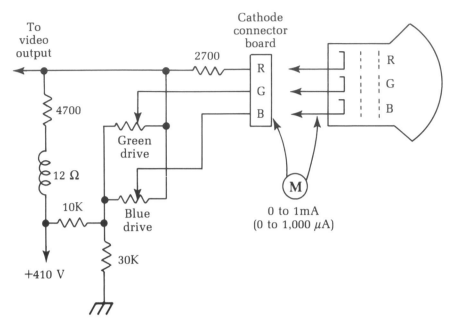

**Fig. 7-6**  *Many color TV sets have provisions for unplugging each cathode lead of the picture tube for testing purposes.*

drive controls and are intended to help the technician get a true black-and-white screen when he's setting up the color temperature. They're used for making the screen color "track" when the brightness is turned up and down, so that it stays black and white. (In color sets, the video or brightness signal is applied to all three cathodes of the color picture tube.)

To check beam current, unhook the cathode of the gun you suspect. Connect a 0 to 1 mA dc meter in series with it. Turn the set on and adjust the screen for average brightness; you should read about 100 $\mu$A (0.1 mA) on each gun, within about 10 percent. If one of the guns has an exhausted cathode, its beam current will be very low, and nothing you can do—such as turning up that screen voltage, etc.—can bring it up to where the other guns are running. Use the currents of the other two guns as a standard; all three should match.

If one gun has a heater-cathode short, you'll probably see all green or all red, whichever one is faulty. Checking the cathode current of the defective gun will probably show you that it is running up to 1 or 1.5 mA instead of the normal 100 $\mu$A. The brightness controls will have no effect. Measuring the cathode-grid voltage will probably show you that this gun has 0 bias. (*Note:* If you read 175 V on the cathode and 175 V on the grid, that gun has 0 bias. The difference voltage between grid and cathode is what you read. Example: cathode has 150 V and grid has 125 V. That means the grid has a −25 V bias on it.)

If this test shows that a gun is bad, verify it by checking the color CRT on a good CRT tester. If you get the same answer, then the gun is probably bad. Considering how expensive color picture tubes are, never take the word of only one test before deciding that a color tube is bad. The trouble might be in the operating voltages; check all dropping resistors, supply voltages, etc., before making up your mind that it's the tube. As said before, the voltages and operation of the other two guns are handy as a standard because all are fed from the same supply in most sets.

## Extension cables for testing TV tuners

TV tuners are simple — only a few circuits. However, they can be hard to get to without the right tools. If you can bring a tuner out where you can get to things, servicing isn't bad at all.

Many tuners today are separate — connected to the chassis by wires, with a coaxial cable to the IF input. You can remove the tuner and leave the chassis in the cabinet. In many cases, however, these wires aren't long enough to leave a console on the floor and have the tuner up on the bench. The solution is to make up a set of extension wires with a terminal board on one end, as shown in Fig. 7-7. Use different colored wires so that you can easily keep

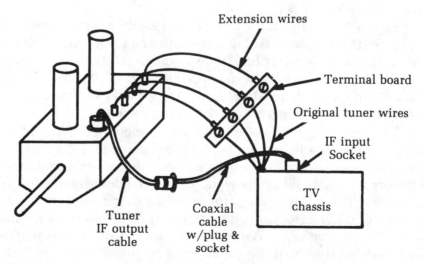

**Fig. 7-7** *You can easily make an extension cable to test TV tuners.*

them straight. If the tuner wires are soldered, disconnect them one at a time and tack the extension wires in their place, fastening the ends of the original wires under the terminal screws. If the tuner has a plug-in cable, you can make up a plug-and-socket extension cable for about 50 cents.

The IF output cable from the tuner is often a plug-in type, with plugs like those used on phonographs. Make up an extension for it, too, with a lug and socket. Any kind of small coax will do because you're only adding a very few picofarads of shunt capacitance. The picture and sound come through in surprisingly good shape.

You can make a rack to hold the tuner while it's on the bench, or just block it up with a couple of empty boxes. Now you can test alignment, make voltage measurements, and make gain checks, etc. a lot more easily than before. Except for the IF output, you can set up all the alignment adjustments of the tuner while the tuner is on the extension cables. You can adjust the IF output easily after the tuner is back on the original cable.

# A quick tuner gain check

Here's a test that will help you isolate that troublesome stage in a tuner or clear a stage of suspicion. You can use a signal generator, but your bench TV antenna works just as well. If there is too much snow in the picture, it is usually a case of tuner trouble. The only question is, "Which stage?" By using this signal-tracing method, you can pin it down in a hurry.

Figure 7-8 shows the signal path of a fairly typical cascode tuner. Starting at point A, hook the antenna to the input terminals. If the picture is very bad (snowy) and the station is known to give a good, clear picture on a normal set, there's trouble. Using the antenna as a signal source, make gain checks through the tuner to find out which stage isn't doing its job.

The antenna has a 300 $\Omega$ balanced impedance. The output impedance of the balun coil is 75 $\Omega$, unbalanced. However, by grounding one side of the lead-in, use the other side and get a satisfactory match for what you want. Put a capacitor of about 100 pF in series with the other side of the lead-in, just in case dc blocking is needed to keep from shorting out bias or dc voltages. In some tube circuits, you can leave out this capacitor, but it is essential for transistor tuners. If the capacitor is not used in a solid-state tuner, the bias on the RF transistors could have damaging results.

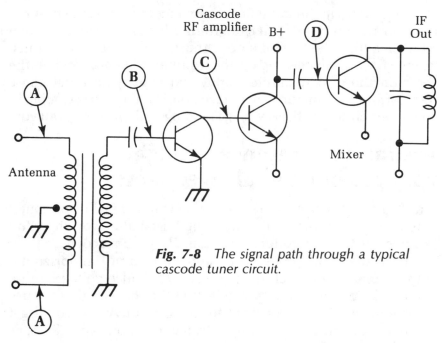

**Fig. 7-8**  *The signal path through a typical cascode tuner circuit.*

To be on the safe side, always use a small dc-blocking capacitor in this type of test.

First, touch the free end of the capacitor to point B, the output of the balun. If the picture clears up, the balun coil is bad. If there is no improvement, keep on going. If you get a better picture by touching down at point C, the first half of the cascode circuit is bad. Change the tube and check the dc voltages. If this shows no improvement, then go on to point D, the mixer grid. If you get a better signal by touching here, then the RF amplifier is faulty.

# Checking for operation of the 3.58-MHz color oscillator

For a loss of color or color sync, one of the first things to find out is if the 3.58 MHz oscillator is running. In the circuit used for many years in RCA sets, this procedure is not difficult. Feed the burst and 3.58 MHz signals into a phase detector. This stage is exactly like the common horizontal-afc phase detector or an FM ratio-detector circuit.

To check the oscillator activity, measure the dc voltage developed across one of the phase-detector diodes. In normal operation the two develop equal voltages of opposite polarity, so you can read either one.

To eliminate the influence of any incoming signal, short the input of the burst-amplifier to ground. If the voltage isn't high enough, adjust the bottom core of the 3.58 MHz oscillator transformer for a peak reading on the voltmeter. Then, remove the short on the burst-amplifier input and adjust the burst-phase transformer for a maximum voltage reading. If there is trouble in any of this circuitry, these voltage readings will promptly tell you.

# Testing for presence of high voltage, horizontal sweep, etc.

The TV screen is dark; is there any high voltage? Is the horizontal output tube working? Is the horizontal oscillator working? You can answer these questions very quickly with a scope.

Hold the scope probe near the plate lead of the horizontal output tube. Don't touch the plate cap; the pulse voltages there will break down the input capacitors of the scope and can cause other damage. You don't need to touch it, anyway. You'll get plenty of pattern height from the tremendous pulse voltages, as you can see in Fig. 7-9.

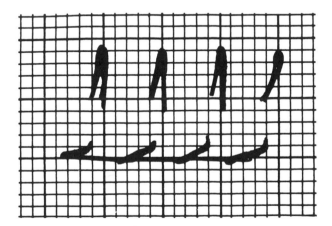

**Fig. 7-9**   *Even without touching the plate lead of the horizontal output tube, a scope will still display plenty of pattern height.*

Check the horizontal oscillator frequency at this point by using the frequency-set method outlined earlier. If you see too many or two few cycles, the horizontal oscillator might be so far off frequency that the output stage will not work.

To check the high-voltage rectifier tube, hold the probe near its plate lead, or even near the bulb. You'll see the same pattern, but with much higher spikes since the pulse voltage is very high at this point if the output tube and flyback are okay. Holding the probe near the high-voltage rectifier output lead to the picture tube also will show you these spikes if the filtering is okay. A pattern with small waves on the horizontal parts, however, as in Fig. 7-10, indicates a severe ringing in the flyback or yoke. In a few cases, an open high-voltage filter capacitor can cause this symptom. Also, trouble in the horizontal-linearity coil or capacitors, or a bad balancing capacitor in the horizontal yoke, can make this kind of pattern.

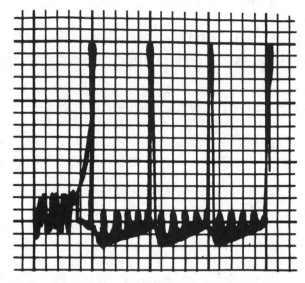

**Fig. 7-10**    *This oscilloscope pattern indicates severe ringing in the flyback or yoke.*

If you see high spikes when you hold the probe near the high-voltage rectifier plate, but none on the dc output lead to the picture tube, the high-voltage rectifier is very apt to be dead. There usually will be good-sized spikes on this lead, even though its voltage is supposed to be filtered.

# Checking for high voltage with a neon tester

A neon lamp will glow in a strong electric field. Such a field exists around the flyback and the plate leads of both the horizontal output tube and the high-voltage rectifier in all TV sets. Special high-voltage testers are made in the form of a rod of insulating material with a neon lamp in a clear plastic housing on one end. If you hold one of these lamps near the plate lead of the high-voltage rectifier and it glows brightly, you know that the horizontal output tube and flyback are working. There is definitely a good deal of energy around there. If there is no dc high voltage, the high-voltage rectifier tube is probably bad.

A neon lamp also will serve as an indicator of the presence of energy around the plate lead of the horizontal output tube, although the lamp won't glow as brightly as it will when near the high-voltage rectifier plate. The RF field here isn't quite as strong, but there will be a definite glow.

# Setting a horizontal oscillator on frequency by comparing with video

You'll find many TV sets in which the horizontal oscillator is obviously off frequency. In some, the oscillator can be thrown so far off by misadjusted controls that you lose the raster, boost voltage, etc. (Children and unqualified technicians are sometimes to blame.) You need a quick way to check the frequency of the oscillator.

Once again, you will use a comparison method. The standard is the horizontal-sync pulses of the video signal. Hook up the scope with a low-capacitance probe and pick up a signal at the video-amplifier plate. Adjust the scope sweep to about 7,875 Hz until you see two horizontal sync pulses on the screen, as illustrated in Fig. 7-11. Set the frequency control and sync lock of the scope to hold them as steady as possible.

Without touching the sweep controls of the scope, move the probe over to the horizontal-oscillator circuit. The grid of the horizontal output tube is a good place to pick up a signal. You can adjust the vertical-gain controls to keep the pattern on the screen, if necessary. Without touching the scope sweep controls, adjust the horizontal-hold control, coils, etc., until you get two cycles of the horizontal frequency on the scope screen, as shown in Fig. 7-12. If misadjustment was the only trouble, your raster will come back and you can use the TV screen to make the final adjustment.

# Testing frequency in vertical-oscillator circuits

If you need to test the frequency of the vertical oscillator and see no picture for some reason (no high voltage, no picture tube, etc.),

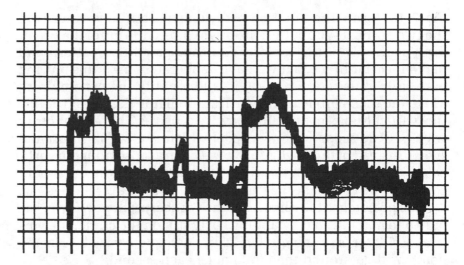

**Fig. 7-11**   *To set the horizontal oscillator frequency, first adjust the oscilloscope to see two horizontal sync pulses on the screen.*

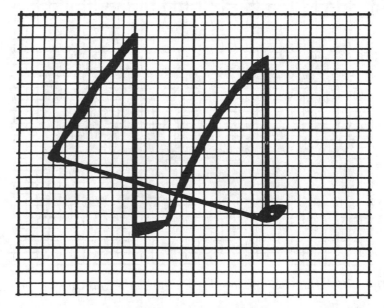

**Fig. 7-12**   *Without touching the scope's sweep controls, make the necessary adjustments to get two cycles of the horizontal frequency on the screen.*

you can make the test with a scope. Compare the oscillator signal with the video signal used to test horizontal-oscillator frequency. Set the scope frequency to show two or three cycles of video signal, near 30 Hz. You can now see the vertical sync pulses, as in

Fig. 7-13. Again, the pattern is badly blurred, but all you need are the two vertical sync pulses: the bright pips at the top of the waveform at the right and center. (You can spread this pattern and make it easier to see, but we have deliberately used this blurred pattern to illustrate how easy it is to identify the sync pulses.) Move the scope probe to the vertical-oscillator circuit and adjust the vertical-hold control to show two cycles. There is another way. Set the scope for line sweep, which is a 60 Hz sinusoidal sweep instead of the sawtooth. Now feed in a pulse from the vertical-oscillator circuit. By juggling the vertical- and horizontal-gain controls, you can make the trace an oval, or even a circle (See Fig. 7-14.) The notch is the spike from the vertical oscillator. If you see only one notch, and it's standing fairly still on the circle, the vertical oscillator is running at 60 Hz. If the frequency is off 1 Hz, the notch will go around the circle once each second.

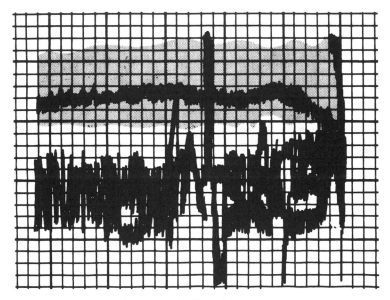

**Fig. 7-13**   *This oscilloscope display shows the vertical sync pulses.*

This procedure is often needed in transistor TV sets that have several ailments at the same time. For example, in one brand, if the vertical circuits are out of adjustment, the vertical-output stage will draw such a heavy current that it will kick the circuit breaker, causing you to waste time looking for nonexistent shorts. Setting the vertical oscillator on frequency will eliminate this problem, even if you have no high voltage and can't see a raster.

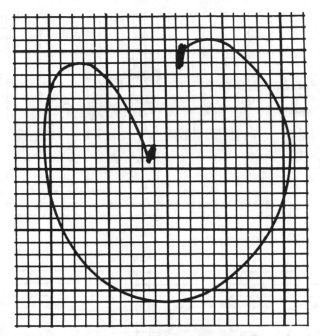

**Fig. 7-14**    *By juggling the vertical and horizontal gain controls, the trace can be made an oval or a circle.*

# Finding the cause of sync clipping with the scope

If you find a TV with a case of sync clipping, or sync trouble of any kind, there are several possible causes. The only quick way to find the trouble is to separate the various possibilities with a scope.

To begin, check the video signal at the sync takeoff point. You can use the direct probe on the scope since this is a high-level signal. You should see a normal video signal, as shown in Fig. 7-15. The sync tips form 25 percent of the total height of the pattern, and the video forms the other 75 percent. If you find such a pattern, the video detector and video output are okay; the trouble is actually in the sync-separator circuits.

If, however, you find something that resembles the trace shown in Fig. 7-16, don't go to the sync separators yet. There's trouble before that point. Notice that here the sync tips are so compressed that they are not actually visible. The video is about normal, but there is little sync. This kind of trouble is usually caused by a bad tube in the tuner or IF stages, incorrect voltages on the IFs, incorrect agc voltages, or even a bad video-detector diode.

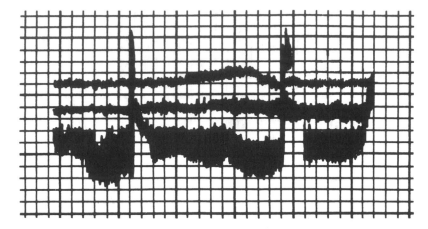

**Fig. 7-15**    *A normal video signal.*

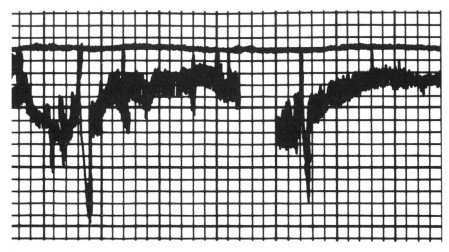

**Fig. 7-16**    *This trace indicates problems ahead of the sync separator circuits.*

The symptom in the signal shown in Fig. 7-17 is the exact opposite. It is referred to as *white compression*. This signal is also a clipped waveform, but notice that the sync tips are normal in height. It's the *video* that is being clipped. The result is a nice, clean raster (no snow) with the picture signals visible only as vague outlines on a very bright background. In bad cases, you usually have to turn the brightness far above normal to see anything.

The most common cause of this symptom is a video output tube that is very weak or gassy, or has heavy grid emission. In some cases, a bad video-detector diode will do it; you'll have to go

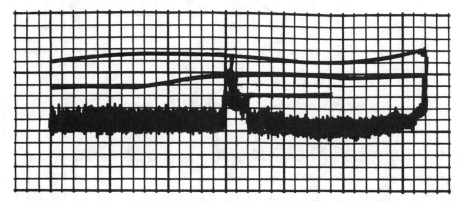

**Fig. 7-17**  *This trace indicates a problem known as* white compression.

by the scope patterns on the input and output of the video amplifier. If the signal is clipped on the video-amplifier grid, the trouble is being fed into the tube and does not originate there. If you do suspect the video amplifier, replace the tube first, then check for correct operating voltages, especially grid bias.

White compression is sometimes confused with a weak picture tube because of the loss of contrast. However, there is one basic difference: if white compression is the trouble, the picture tube will still be able to make a very bright, well-focused raster. If the CRT is weak, you'll have very low brightness, and probably loss of focus, especially in highlights. With white compression, highlights smear out but are very bright. If you get about 50 V p-p of good clean video at the grid or cathode of the CRT but still have a poor picture, then suspect the picture tube.

## Finding sync troubles with the scope

When you encounter sync troubles, there's only one instrument that will tell you anything useful: the scope. You can find all sync trouble in the least time by following the sync signal from the point of origin through the various sync-separator and amplifier stages until you find the point where it disappears. Voltage and resistance checks will then pinpoint the cause of the trouble.

Because most of these are high-impedance circuits, use the low-capacitance probe to keep from disturbing them too much. Start at the sync takeoff point, which is usually in the plate circuit of the video-output stage. Make sure the composite video signal is there — with plenty of sync — before you go any further. The signal should look like the one shown in Fig. 7-18. You might see

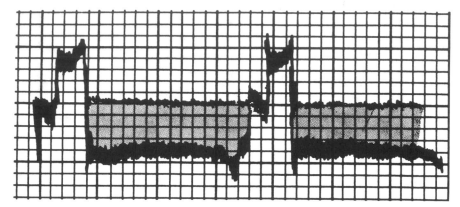

**Fig. 7-18**  *The composite video signal should contain strong sync pulses.*

signs of video compression, and in fact, the signal in the figure shows a sync-to-video ratio that is closer to 50:50 than to the 25:75 ratio you look for in the full video signal. However, this waveform was taken off at a point lower in the video-output circuit than the full video signal applied to the picture tube.

The correct scope patterns for various points in the sync circuit usually are shown in the service data or on the schematic diagram. For instance, the stripped signal might be similar to Fig. 7-19 after the video has been clipped. This is the type of pattern seen at the input to the vertical integrator in many sets. The sharp spikes are vertical sync; the rest are horizontal sync pulses.

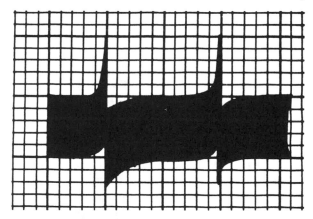

**Fig. 7-19**  *After the video has been clipped, the stripped signal might look like this.*

## Vertical sync

Vertical-oscillator circuits depend mainly on the amplitude of the sync for proper operation. Horizontal sync works mostly on phase and will lock in on a weak signal far longer than will the vertical. So, any weak-sync condition will show up as vertical-roll troubles first.

To check for sync amplitude, hook the scope to the vertical oscillator circuit. Figure 7-20 shows a typical pattern on an oscillator grid, and Fig. 7-21 shows one as seen on a plate. If you turn the vertical-hold control so that the blanking bar rolls slowly down, the vertical-sync pulses appear as pips on the waveforms. Although some vertical oscillators lock satisfactorily on less sync than this, it's always nice to have a sync pulse of this amplitude for good, tight-locking action.

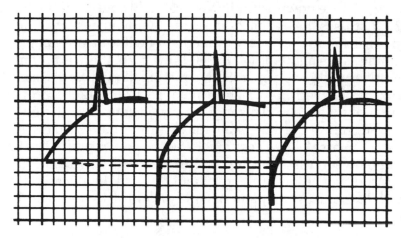

**Fig. 7-20**    *This pattern should be found at the base (or grid) of a vertical oscillator circuit.*

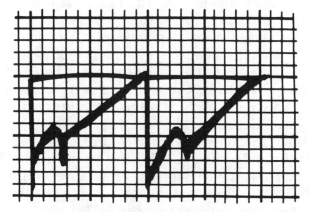

**Fig. 7-21**    *This pattern should be found at the collector (or plate) of a vertical oscillator circuit.*

Ordinarily, you just look at this waveform to get a good idea of the sync amplitude (in proportion to the amplitude of the full waveform). If you want, read the sync-pulse amplitude by itself. Kill the vertical oscillator and take a reading of the sync at either the grid or the plate of the vertical oscillator — depending on where the sync-injection point is in the circuit. In all cases, it will be found at the output of the vertical integrator, coming from the last sync-separator or sync-amplifier stage. The amplitude of sync for the set getting tested is usually given on the schematic. In most circuits, sync fed to a grid is positive-going, and to a plate, negative-going.

You can tell if there is any sync in the vertical circuits by rolling the picture down with the vertical hold control. If the blanking bar crosses the bottom of the screen with no hesitation, and if you can also roll the picture up smoothly without any stopping or jumping as the blanking bar leaves the screen, there is no vertical sync. Normally, as you roll the picture down, you'll see the blanking bar snap out of sight when it gets 2 or 3 inches from the bottom of the screen. This indicates good sync-lock. If you turn the hold control the other way, the picture should hold to a given point, then break loose and travel upward very rapidly. To distinguish between the two, most technicians call downward movement *rolling* and upward movement *flipping*. In all cases, if you can roll a picture slowly and smoothly upward, there is sync trouble.

## Horizontal sync and afc checking

Horizontal sync circuits are just as easy to check as vertical circuits. In many sync circuits, you'll find composite sync at the output of the sync separator: Vertical and horizontal sync pulses are separated by resistor-capacitor networks. The low-frequency vertical sync goes through large capacitors, with good-sized bypass capacitors that shunt the higher frequency horizontal sync to ground. The horizontal sync goes through very small capacitors, which have a high reactance to the low-frequency vertical pulses.

The typical horizontal afc circuit is a phase comparer. The horizontal sync from the TV signal is compared in phase to a pulse from the horizontal oscillator. This pulse sometimes is taken from the oscillator itself, sometimes from a special winding on the flyback. Both signals usually are shaped into a sawtooth waveform by resistor-capacitor networks. Figure 7-22 shows a typical sawtooth found on an afc diode plate.

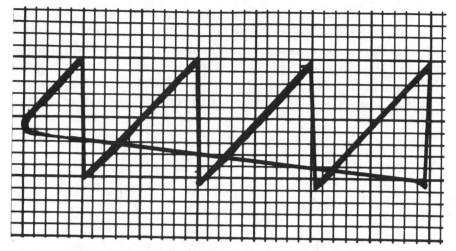

**Fig. 7-22** *A typical sawtooth wave signal found at the agc diode anode (or plate).*

Check for the presence of both sawtooth waveforms and compare their amplitudes to the values given on the schematic. As noted, horizontal sync works on phase, so the amplitude isn't too crucial. However, it must have a certain minimum value before the afc circuit will work properly.

The easiest way to check an afc circuit is by the process of elimination. Take it out of the circuit by shunting the horizontal oscillator grid, etc. Then see if the oscillator will work alone. If so, it will make a single, floating picture on the screen. Now put the afc back; the picture should lock in and stay in sync. If replacing the afc makes the picture go out of sync, or if the picture continues to float (no sync), then the afc circuit must be bad.

If the afc uses a pair of diodes, check them—preferably by substitution, but you can check them for front-to-back resistance ratio, or shorts, or opens, etc. If the diodes are okay, then check the little sync-coupling capacitors, the 82 to 100 pF micas. These capacitors have been known to leak; if they do, they could upset the dc balance of the afc circuit. Also, check all resistors and capacitors. There are only six or eight of them in an afc circuit, and they must all be good if the set is to work properly.

# TV signal tracing and gain checks with an RF or AF signal generator

You can check out a video-amplifier stage, video-detector stage, or video IF stage with an RF or AF signal generator. The audio output from the RF signal generator can also be used if you don't have a separate AF generator. Figure 7-23 illustrates the pattern when a square-wave audio signal (about 600 Hz) is fed into the input of the video amplifier of a black-and-white TV set. Notice the sharp horizontal bars on the screen. (You also can use this test to adjust the vertical linearity. Note the compression near the top of the screen.)

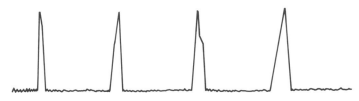

**Fig. 7-23**   *A typical keying pulse.*

The contrast of the bars is used to check out the gain of the video amplifier. The input signal level should be 2 to 3 V p-p in single-stage video amplifiers like this one. In color TV video-amplifier stages, check the schematic for the average level of video output. If a low-level signal is needed and the attenuator of the signal generator won't go down that far, you can make up a simple resistive voltage divider as described earlier. Don't over-load the input, especially in the video stages of transistor TVs.

A scope pattern of the signal in Fig. 7-24 is shown in Fig. 7-25. Notice that there is some distortion of the square wave. If a video amplifier has a very wideband output and if test conditions are just right, you see an almost perfect square wave — up to about 10 kHz, with sharp, clean bars on the CRT. However, the set used for this test produced a very acceptable picture, so don't be overly critical.

You can often improve the scope pattern at the output by tak-ing off the picture-tube socket and picking up the signal at the grid or cathode pin of the socket. This method substitutes the input capacitance of the scope for that of the picture tube and sharpens the pattern. The pattern of Fig. 7-25 was made with the CRT still hooked up.

A sine-wave signal can also be used. In fact, you can feed an amplitude-modulated RF signal into the input of the set's video IF stages and see the resultant on the screen, as in Fig. 7-26. (The contrast of the bars also gives you a good idea of the condition of the picture tube.) This is a valuable shortcut test when the big

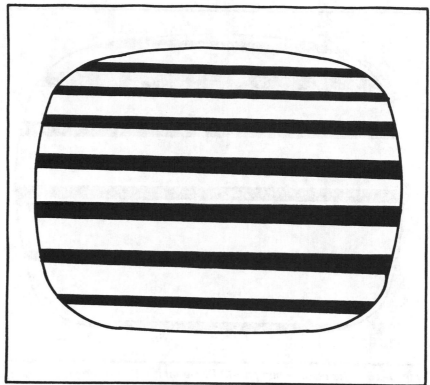

**Fig. 7-24**  *The pattern when a 600 Hz square wave signal is fed to the input of the video amplifier.*

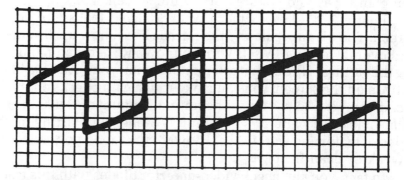

**Fig. 7-25**   *The oscilloscope pattern for the signal in Fig. 7-24.*

question is whether the trouble is in the video IF or in the tuner. If an RF signal at the picture IF frequency (modulated by a 400-Hz sine wave) passes through the IF stages, video detector, and video amplifier and makes a pattern on the screen, then you can be

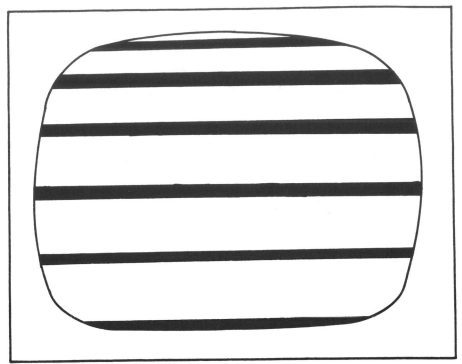

**Fig. 7-26**   *The result when an AM RF signal is fed into the video IF stages.*

fairly sure the trouble is in the tuner, especially if the original complaint was "no picture" or "white screen."

Figure 7-27 shows what a weaker signal looks like on the screen. The bars are pale and the retrace lines show up. By checking several working TV sets with your own RF generator and noting the setting of the output attenuators when you get a good black pattern on the screen, you can make rough gain checks on the IF, etc. If you have to use twice the normal amount of signal to get a good black-bar pattern, something is weak. By starting at the video detector and working back toward the tuner, you can locate a weak or dead IF stage.

In fact, you can check video-detector diodes with this test. If an IF signal fed into the input produces very pale bars, but an audio signal fed into the video-amplifier grid produces good, sharp bars with plenty of contrast, then the video-detector diode could be leaky or open. When making this test, check the setting of the agc control. If the agc is too negative, it will cut the IF gain and make the picture weak and washed out.

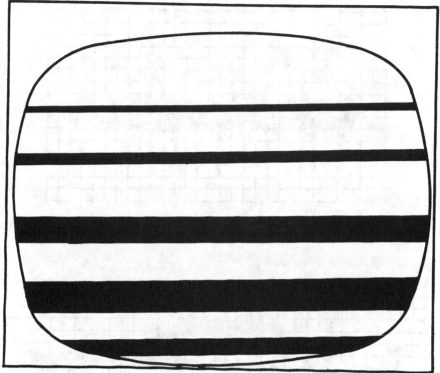

**Fig. 7-27** *A weaker signal looks something like this.*

# Signal tracing with the scope and color-bar patterns

If you run into trouble in the video IF or video-output stages, signal-trace through these circuits with the scope to find where the signal stops or loses gain. If you have an easily identified pattern, the job will be easier. Such a pattern can be supplied by the signal from a bar RF/IF dot generator set for the color-bar signal or the crosshatch. In circuits where the signal is still RF, such as the video IF, you need a crystal-detector probe on the scope. After detection, you can use the direct probe, although a low-capacitance probe is handy in many cases. The main benefit in using this signal is that it can be readily identified.

Figure 7-28 shows a crosshatch signal, taken from a video amplifier through a low-capacitance probe. Figure 7-29 shows a color-bar signal at the same point, taken with a direct probe. The square pulse near the right is the horizontal sync bar of the signal.

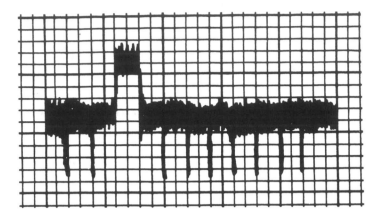

**Fig. 7-28**    *A crosshatch signal taken from a video amplifier through a low-capacitance probe.*

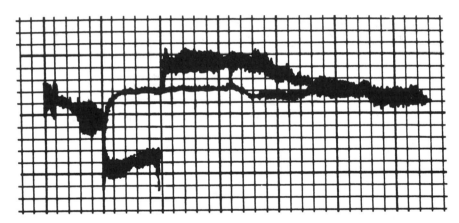

**Fig. 7-29**    *A crosshatch signal taken from a video amplifier through a direct probe.*

The set used for these photos was a black-and-white TV, so the color-bar signals are not as plain as they are in color sets. Nevertheless, the characteristic shape of the signals can be seen. Most bar, dot, and crosshatch signals, displayed on the scope at a vertical rate of 30 Hz, resemble a comb. Check them out on a set that is working so you'll know what they should look like. So far, be mainly concerned with amplitude rather than waveform. Don't be too critical of any distortion, at least not yet. However, a test pattern can be useful for detecting sync clipping. An example is illustrated in Fig. 7-30.

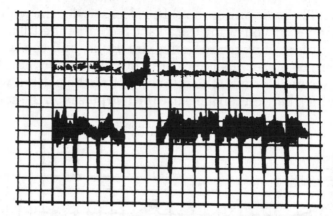

**Fig. 7-30**   *This test pattern is useful for detecting sync clipping.*

# Using the scope to set the duty cycle of a horizontal-output transistor

Correct adjustment of the oscillator-drive waveform in a transistor TV is not as simple as in tube types. The wrong adjustment can cause trouble in a hurry, even faster than in tube sets.

The horizontal-output transistor actually works as a switch; it is driven by a rectangular pulse waveform. In some sets, you must use the scope to set the ratio of off-time to on-time (duty cycle) in this waveform.

If the on-time is too long, average current will go up, which can cause overheating of the junction and even blow the transistor or kick out the circuit breaker. By using the scope to display the waveform, measure the ratio between on-time and off-time directly.

The complete procedure is similar to the procedure used in the old, faithful synchroguide circuit. Two stabilizing coils are used. One controls the oscillator off-time, and the other (which is actually resonated at about 40 kHz) controls the on-time. To make the preliminary setup, ground the collector of the sync-separator transistor and hook a jumper across the sine-wave coil. Now, turn the set on and adjust the hold control for the most stationary picture. The picture floats since there's no sync or stabilization, but you can get a single picture by juggling the hold control.

The bottom core of the coil (the 40-kHz section) is adjusted for a pulse-width ratio of 1 : 2 (1 on, 2 off). The waveform is shown in

Fig. 7-31. The wide pulses should be at least twice the width of the narrow ones. If you want the exact recommended figures, the narrow pulses (downward-going) are 18 $\mu$s, and the wider off-pulses are 60.5 $\mu$sec. As long as the ratio is greater than 1:2, it's fine. You can determine the ratio by setting the pattern width on the scope so that you can measure the two pulses on the calibrated screen.

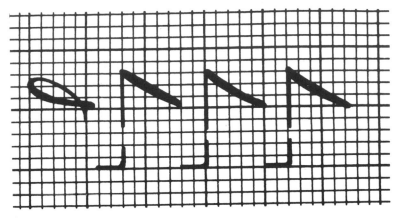

**Fig. 7-31**   *The bottom core of the coil is adjusted for a pulse width ratio (or duty cycle) of 1:2.*

Next, turn off the set and remove the jumper across the sine-wave coil. (Do not connect or disconnect anything in a transistor circuit with the power on. It's dangerous. Making or breaking a connection could cause transients that would puncture transistors. Turn the set off!) Turn it back on and adjust the top core — the sine-wave coil — for a locked-in picture. Remove the scope and the job is done. Disconnect the jumper that grounds out the sync, too, before you forget it.

# How not to read voltages:
# The "do-not-measure" points in TV

Now that you know how to read dc voltages at typical points of a TV circuit, let's see about those points that must not be measured: the plate of the horizontal output tube, the high-voltage rectifier plate, the damper tube, and the vertical output plate. These points should be marked DO NOT MEASURE on the schematic.

They say do not measure because there is a very high *pulse* or *spike* voltage present at all times, in addition to the dc plate voltage. During normal operation, these spikes reach 15,000 to 20,000

V peak. Such spikes of voltage can damage the multiplier resistors in your VOM by causing a flashover (internal arcing) in the precision resistors. These resistors are seldom rated at more than 1 W —usually less. Therefore, VOMs should never be subjected to this kind of mistreatment. Once a multiplier has flashed over, the meter will be inaccurate until it is recalibrated.

You can read the dc voltage on these TV circuit points if you use the right methods. In each case; the voltage is fed through a coil—the horizontal or vertical output transformer, etc. So, turn the set off and read the resistance of the coil primary. Check this value against the value given on the schematic. Now, turn the set on and read the dc voltage at the bottom of the coil—the opposite end from the plate of the tube. This is where the supply voltage is fed in. If it's okay there, and if the coil winding shows continuity, there is plate voltage on the tube in question.

# Blown audio output transistors

Audio output transistors seem particularly susceptible to blowing out. Many technicians have run into problems where they replace a blown audio output transistor and the new one immediately blows out when power is applied. This clearly indicates that the blown transistor is just a symptom of some other problem.

How can you locate the problem without being able to take any voltage or current measurements? The solution is to use a Variac. Before applying power, plug the set into the Variac and adjust its output to 0. Now turn on the TV and monitor the current flowing through the audio output transistor as you slowly increase the voltage from the Variac. Many schematics indicate how much current should be flowing through this transistor.

Generally, TV audio stages operate in the class-B mode. In this type of circuit, the quiescent (no-input signal) current drain should be considerably less than the full-load current drain. If you detect any appreciable current drain with a low supply voltage, trouble is clearly indicated.

# Dealing with dead sets

To the average layman, the ultimate equipment catastrophe is a completely dead set—no raster, no sound, no lighted indicators, no nothing. However, in most cases, a completely dead set is the easiest to service. It is certainly less frustrating to work on than one with an intermittent problem.

When you are faced with a completely dead set, the first step is to check the painfully obvious. Is the set plugged in? You might be surprised to learn how many service calls are made for unplugged equipment.

Even if the set is plugged in, that is no guarantee that the power is being applied to the set. Is the socket live? Is there a blown power line fuse? Sometimes just a single socket goes dead, so always try plugging something else into the questionable socket before drawing any conclusions, for example a lamp. Many technicians carry electrical socket testers in their tool kits. These devices are handy and quick. In addition to giving a clear indication of power, they also indicate whether or not the socket is properly grounded. These simple testers only cost a few dollars.

If power is being applied to the set, check the power cord. There could be a small break in it. Many sets have detachable power cords. Make sure the cord is fitted into the set properly and securely.

Better quality sets normally have internal fuses, or more commonly, circuit breakers. If the fuse is blown, obviously it should be replaced. If the circuit breaker is tripped, reset it. Then let the set run for a while. The problem could have been caused by a brief transient in the power lines, but there could be something wrong within the set's circuitry. Generally speaking, if the problem is still present, the fuse or circuit breaker will soon blow again. In some cases it blows instantly as soon as power is applied.

Serving this kind of problem can be rather tricky. After all, you can't take any voltage or current measurements. Do not try bypassing the fuse or circuit breaker. Remember, it is blowing for a reason. Without the current protection of the fuse or circuit breaker, other possibly expensive components could be damaged or even destroyed. There might be a risk of fire hazard.

Try using your nose and eyes. Does something in the circuit smell or look burned? Especially look for discolored resistors that might have changed values. If anything looks even remotely suspicious, take a passive resistance reading. You might have to lift one end of the questionable component from the circuit to get an accurate or meaningful reading.

Check the wiring closely for any potential short circuits. Once again, the ohmmeter can be useful to eliminate any doubt. Watch out for frayed insulation on any jumper wires or bare leads that might have gotten bent from their original position. Occasionally, a small glob of solder can break free and rattle about, eventually shorting something out.

Initially, it makes sense to confine your suspicions to the power supply itself. To determine if the power supply is in fact the culprit, disconnect it from the set's other stages. Apply power to the supply circuit without any load. If the fuse or circuit breaker blows, the power supply is definitely at fault.

If the power supply seems to work okay without any load, you might want to try it with a dummy load. Sometimes a power supply will work fine unloaded, but the fuse/circuit breaker will blow when the supply attempts to drive a load.

If the power supply seems to work fine when disconnected from the rest of the set, try reconnecting other stages *one at a time.* When the fuse/circuit breaker blows, the troublesome circuit is on line.

When the problem is in the power supply itself, which it will be more often than not, a good test procedure is to unplug the set and discharge all the large capacitors. Then measure resistance from various points in the power supply circuit to ground. You should find fairly high resistances to ground throughout the circuit. In most cases, the resistance should be at least 50K, if not more. A resistance reading of 0, or close to it, almost surely indicates a problem spot. A moderate reading—say, about 15K to 25K, or so—indicates that you are close to the trouble spot, but haven't quite pin-pointed it yet. Shorted capacitors are common causes for such problems.

If some sets, the power supply will be working fine, but the set is still completely dead, except perhaps for a power indicator light. This can certainly be a frustrating problem, but at least you can take voltage measurements. In this circumstance, dc voltages generally will be the most useful. If you find a voltage that is significantly incorrect or missing altogether, a problem is definitely indicated somewhere in that stage of the circuit. Find and correct the problem(s) as in any other repair job.

Work your way through the set's stages from input to output until you turn up the trouble. If the dc voltage tests are unrevealing, go back to the beginning and try ac signal tracing.

# 8
# Special tests

THIS CHAPTER INCLUDES A VARIETY OF UNIQUE TESTS THAT REQUIRE special techniques with common equipment.

## Measuring peak voltages without a voltmeter

Frequently, you need to know the peak to peak drive signal voltage on horizontal output tubes, vertical output tubes, oscillators, etc. Normally, you read this voltage on a calibrated scope or with a peak to peak reading ac voltmeter. If these instruments aren't available, however, you can always use the low-milliampere range or the microampere range of the VOM.

Open the bottom end of the grid resistor and hook the microammeter in series with it. Connect the positive terminal of the meter to ground and the negative terminal to the resistor. With the circuit in operation, multiply the current you read by the value of the grid resistor. The result, according to Ohm's law, is the peak value of the grid voltage. Note that we said *peak*, not *peak to peak*; the grid conducts current only on the positive-going halves of the drive signal. So, to get the peak to peak value most often specified in service data, double the value calculated.

In one actual test, a 6DQ6 tube with a 470K grid resistor read a bit over 100 $\mu$A. This gives 47 V *peak*; the *signal*, measured with a calibrated scope, was about 90 V peak to peak. Resistors in such circuits aren't precise; tolerances are 10 to 20 percent. Voltages obtained by this method won't be exact, but they will be accurate enough to give the information needed pertaining to the drive signal voltage. Remember to double the reading for a peak to peak value.

# Using a pilot lamp for current testing

When you encounter one of those jobs that runs along fine then suddenly blows the fuse, it can be expensive as well as annoying. Fuses cost money, especially if you blow four or five of them before you find the short. Make up a test adapter with a pilot-light socket wired across a blown fuse, as shown in Fig. 8-1. Use a pilot lamp that has a rated current a little higher than the normal drain of the circuit.

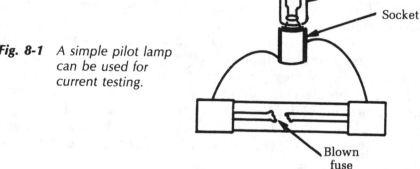

***Fig. 8-1*** *A simple pilot lamp can be used for current testing.*

For instance, if you're checking the B+ supply of a TV set rated at 200 mA, use a 250 mA pilot lamp—No. 44. There are pilot lamps with current ratings all the way from 150 mA (No. 40, No. 47) to 0.5 amp (No. 41, etc.). There also are special lamps rated as low as 60 mA (No. 49) if you happen to need them.

When you turn the set on, the pilot lamp will glow a medium yellow. You can watch this while you tap, move, heat, or test parts in the circuit. If you hit something that is causing the short, you'll see the lamp flare up to a bright blue-white. A dead short will blow the lamp out, of course, but pilot bulbs are usually cheaper than slow-blow fuses!

As a matter of fact, Motorola and other two-way radio manufacturers have used pilot lights as B+ fuses in power supply for several years. If you run into one of these circuits, be sure to use the same type of lamp as a replacement; it must have the correct current rating.

# A voltage divider for obtaining very small audio signals

For audio-amplifier testing, you need a source of low-level signals. Be very careful not to overload the input, in transistor amplifiers especially. Also, you can make a quick check on any audio amplifier by feeding in a given number of millivolts and measuring the audio-output power. If the amplifier comes up to specifications, there's no need to go any further. This is also a fast way to isolate a weak channel on a stereo system.

You can read the output of an amplifier in watts by substituting a load resistor of the proper value for the speaker, applying a steady signal to the amplifier input, and measuring the voltage across the load resistor. The value of power output can be figured out by the Ohm's law equation $W = E^2/R$. In many cases, the right values of voltage for both input and output are given in the service data.

For instance, an amplifier might call for 2.75 V across the output-load resistor at an input of 300 mV. The problem is to get a true 300 mV since most shop-type audio generators don't have accurately calibrated attenuators. To get the signal needed, use a simple resistive voltage divider, as shown in Fig. 8-2. If the upper resistor has a value of 700 Ω and the bottom resistor is 300 Ω and you apply exactly 1 V of audio signal, you can take off 300 mV at the tap.

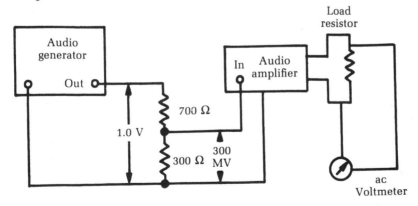

**Fig. 8-2**   *A simple resistive voltage divider can be used to bring the signal down to the desired level.*

It might be more convenient to use a 1,000 Ω variable resistor, as in Fig. 8-3. Hooked up as a potentiometer, the slider can be set with an ohmmeter to give any value of signal output needed.

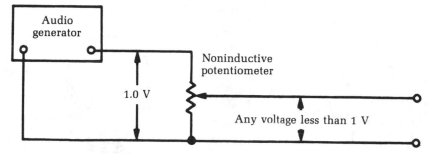

**Fig. 8-3**   *It might be more convenient to use a 1K potentiometer in the voltage divider network.*

Stereo and hi-fi amplifiers designed for low-output magnetic cartridges usually call for about 5 to 10 mV, while "ceramic" inputs call for 700 mV to 1 V. With the signal-generator output set at a given level, the pot can be used to obtain any fraction of this amount. This method eliminates the need for accurate measurement of very low signals; all you need to know is the total signal across the divider and the way the divider is set up. A carbon potentiometer is best. The inductance of a wirewound pot could affect the accuracy of division at high frequencies.

## A quick test for audio power output

There's a quick and dirty test for audio power output if you don't want to bother with calibrated resistors. Get a weatherproof lamp socket and put a couple of heavy terminal lugs on the wires—spade lugs are best. Connect this socket across the high-impedance output taps on the output transformer of the PA system, as illustrated in Fig. 8-4. Screw a standard incandescent lamp into the socket of a wattage to match the power you want to check. For example, on a 50 W amplifier, use a 50 W lamp; for a 30 W amplifier, use a 25 W lamp (the nearest commercial size).

Now fire up the amplifier and feed in an audio signal. If you have the rated power output, the lamp will light. If a 50 W amplifier will light a 50 W lamp to a good white or about normal brilliance, that's it.

The impedance match is closer than you'd think on the 500 $\Omega$ output. A 25 W lamp, for instance, draws about 0.21 amp; plugging this value into the formula $W = I^2R$ gives a hot resistance of about 570 $\Omega$, which is fine for all practical purposes. If you want, you can parallel two 25 W lamps across a 250 $\Omega$ output and get a fairly

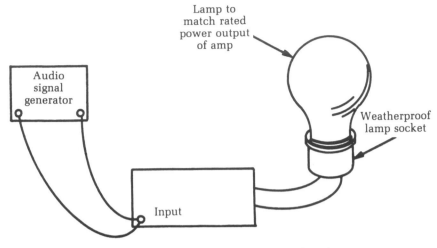

**Fig. 8-4**  *The test hook-up for a simple check of audio power output.*

close match on a 50 W amplifier. This method isn't good for high-power transistor amplifiers because more critical impedance-matching is needed, especially in the output-transformerless types. For them, use the exactly matched load resistor mentioned in the earlier test.

# Measuring amplifier output power with a multimeter

The lamp test for audio amplifier output power described previously is sufficient for some purposes, but it is obviously a very crude and inexact test. If you need a more exact indication of an audio amplifier's output power, you can use an audio wattmeter, but such a device is not a standard element on most electronics workbenches. If you do a lot of work with audio amplifiers, you probably will want to buy a good audio wattmeter, but most technicians will only have an occasional need for such an instrument. Rather than clutter up your work area a lot of equipment that will spend most of its time just gathering dust, you might want to use your general test equipment.

For this test, you will use a signal generator or audio frequency oscillator and a multimeter. It doesn't much matter if the multimeter is of the analog or digital type in this application.

The signal generator can put out almost any simple ac waveform. A sine wave would probably be the best choice, but this is hardly critical. A frequency of 1000 Hz (1 kHz) would be the ideal for performing this test procedure, but again this isn't terribly critical. You don't have to worry about setting the signal frequency precisely.

The signal generator you use should have a low output impedance, of no more than 600 Ω or so. To prevent overloading the inputs of the amplifier being tested, it is highly desirable to use a signal generator with an attenuator (output signal level) control. Set the signal level to suit the specific amplifier's input requirements. For instance, most standard stereo amplifiers call for a 1 V input at the auxiliary input jack(s) to drive the amplifier to its full power output, which is exactly what we want in this test.

The basic test setup is illustrated in Fig. 8-5. Set up the multimeter to read ac volts. Notice that a load resistor is used in place of the loudspeaker at the output of the amplifier to avoid the complexities and possible misreadings that might result from the ac impedance of the speaker's voice coil. Things will be much simpler if only straight dc resistance is involved.

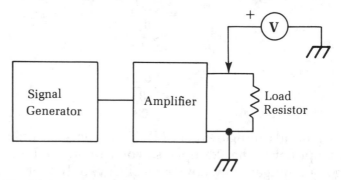

**Fig. 8-5**    *The basic test setup for measuring amplifier output power with a multimeter.*

The resistor's value should be equal to the nominal impedance of the speaker. For most modern stereo amplifiers, this amount will probably be 8 Ω. Other common values you might encounter are 4 Ω and 16 Ω. Some PA amplifiers might call for speakers with higher impedances. In most cases, the correct output impedance value will be indicated on the back of the amplifier near the output jacks or screw terminals.

A high-power resistor must be used in this application. The resistor's power rating should be equal to at least the expected power output of the amplifier being tested. We strongly advise overrating this resistor's power handling capability by at least 10 to 25 percent. For example, if the amplifier you are testing is rated for 10 W, use a resistor that can withstand at least 11 to 13 W.

If the amplifier being tested has any tone controls or filters, set them all for flat response while performing this test procedure.

You can use the ac voltage measured across the load resistor to determine the output power with this standard formula:

$$P = \frac{E^2}{R}$$

where $P$ is the power in watts, $E$ is the measured voltage in watts, and $R$ is the value of the load resistor.

As an example, assume that the load resistor has a value of 8 $\Omega$. Also assume that the amplifier you are testing is rated for 15 W.

It is a good idea to first perform this test at less than full volume. Set the amplifier's volume control at about one-half to two-thirds of full scale. Let's say we measure an ac voltage of 7.5 V. What is the wattage of the amplifier's output signal? Just plug the known values into the equation:

$$P = \frac{7.5^2}{8}$$
$$= \frac{7.5 \times 7.5}{8}$$
$$= \frac{56.25}{8}$$
$$= 7.03125 \text{ W}$$

You can round this figure off to 7 W. This seems reasonable for the 15 W amplifier of the example, so you can proceed. If, however, you got a strangely low power level (say, only 2 or 3 W), or an oddly high wattage (say, 25 W), there is apparently something wrong. Trouble-shoot and correct the problem before proceeding. Attempting to run a defective amplifier at full volume might cause extensive damage, which is why we performed the test at a lower volume first.

Now, crank up the amplifier's volume control all the way to its maximum setting and repeat the measurement. This time, let's say you got a reading of 11.25 V. The maximum power output is:

$$P = \frac{11.25^2}{8}$$
$$= \frac{11.25 \times 11.25}{8}$$
$$= \frac{126.5625}{8}$$
$$= 15.820312 \text{ W}$$

This is close enough to your expected rated value of 15 W. It's a bit high, but not excessively so. The difference might be due to several causes, including accumulated measurement errors, component tolerances, or possibly the manufacturer simply rounded off the actual power rating to give a little "fudge room."

It will be rare to get a too-high power level. If the amplifier is malfunctioning, it will probably give a much lower than normal wattage value. A defective transistor in the amplifier's circuitry is a likely cause.

This test is useful, but it is still fairly crude. It doesn't give you much information. You know the absolute signal amplitude, but you don't know how badly the signal is being distorted by the amplifier or how smooth the frequency response is. The output power might be significantly higher or lower at other signal frequencies. Other tests will be needed to identify such problems.

An oscilloscope can be used to help spot distortion problems. Refer back to chapter 5. A test procedure for an audio amplifier's frequency response is outlined in the next section.

# Testing an audio amplifier's frequency response

No practical amplifier circuit is perfect. All practical electronic circuits are frequency sensitive, at least to some extent. Not all frequencies will be amplified by the exact same amount. Some will receive a little extra boost, while others will be attenuated. A high-fidelity amplifier is designed to exhibit as flat a frequency response as possible. The differences in amplification for all frequencies in the audible spectrum should be nonexistent, or at least reduced to a negligible level. For many modern amplifier circuits, there are no measurable fluctuations in the audible frequency response. There probably are some, but if they are too small to measure, they are almost certainly inaudible.

The ideal frequency response specification for a high-fidelity audio amplifier is 0 dB. This specification indicates a perfectly flat response within the stated frequency range. (If no frequency range is specified, the standard audible range of about 20 Hz to 20 kHz (20,000 Hz) can be assumed.) A practical audio amplifier might have a frequency response specification like 25 W±1 dB. This means that no frequency signal within the audible range will be less than 1 dB below 25 W or above 1 dB above 25 W when the amplifier is operated at full volume.

The decibel (dB) is a comparative value. It has no absolute meaning, like the volt. It is a definite, unambiguous unit. But to say a certain signal is 1 dB is meaningless unless you say what it is being compared to. It is 1 dB above (or possibly below) some reference point. In audio work, you sometimes will encounter dB values with no stated reference. Audio technicians have a standard, agreed upon reference value of 6 mW (0.006 watts), which is equal to 1.9 V across a 600 Ω resistance. For our purposes, if you see that an amplifier has a power level of 5 dB, you know that its power rating is 5 dB greater than 6 mW.

It is important to realize that the dB scale is logarithmic, rather than linear. A value of 2 dB is not twice as much as 1 dB ion the same way that 2 V is twice as much as 1 V. For decibels, a doubling of power is a difference of 6 dB.

Most analog multimeters have a decibel scale, usually the bottommost scale on the dial face. Some deluxe DMMs also might have a dB scale, but they are the exceptions, rather than the rule, so you will probably want to use an analog multimeter for this type of test.

The centerpoint on the dB scale is marked 0. Readings to the right of this point are positive, and readings to the left are negative. A negative dB value indicates that the measured quantity is less than the reference value; a positive dB value indicates that the measured signal is higher than the reference value.

The dB scale on a multimeter is calibrated assuming the 600 Ω audio standard. (Some sources give a standard impedance of 500 Ω.) In practical terms, the difference is negligible except for high-precision tests. If you need that much precision, you should be using specialized laboratory-grade equipment anyway, not a simple VOM.

The values printed on the dB scale are valid only for one specific ac range, usually (though not always) the lowest ac volts range the meter offers. For higher ranges, you will need to add a specific adjustment factor. The number of dB to be added to the reading for each range is usually printed on the meter's faceplate in the lower right corner. For example, a particular multimeter might have the dB scale calibrated for the 5 Vac range. For the 10 Vac range, you must add 6 dB to each reading, and on the 50 Vac range, you must add 20 dB. If you are operating the meter on the 10 Vac range and get a reading of 0 dB, the measured signal has a value of 6 dB [0 db (reading) + 6 db + 6 dB (adjustment factor)].

To actually make the measurements, use the same setup from the output power test described earlier. A signal generator drives the amplifier (or other audio circuit or device), and the reading is taken off the load resistance. This test setup was shown in Fig. 8-5.

For most audio amplifiers you will have a severe impedance mismatch. The amplifier's output is designed to drive an 8 Ω (typical) load, but the multimeter is calibrated assuming a load resistance of 600 Ω. You will not get very accurate readings using this method. Shortly we will give an alternate method. You can get some crude relative readings, which might be sufficient for some quick-and-dirty testing, but the actual dB values will probably be way off.

On the other hand, the output impedance of some audio equipment will have the correct impedance. Such equipment includes tape decks, preamplifiers, mixers, microphones, and synthesizers. You can use this test procedure directly for such devices.

Set all tone controls, equalizers, and so forth to their flat position before making any test measurements. Set the volume control in one position and leave it alone. Typically, the halfway point will be a good choice for basic frequency response tests.

Measure the output signal level using the dB scale with the signal generator set for the lowest frequency of interest. Note the dB reading. Now, raise the signal frequency of the signal generator. Check the input signal voltage to make sure it has not changed with the change in signal frequency. If it has, adjust the signal generator's attenuator so the circuit under test sees the same input signal level. Now, measure the output signal level on the dB scale, as before. Make a note of the value. Repeat this process for a number of discrete frequencies throughout the range of interest.

Ideally, you should get the same dB value for each test frequency across the unit's desired range. Any fluctuations in the values indicate irregularities in the device's frequency response. It is often helpful to graph the results for a visual image of the circuit's overall frequency response.

A good set of test frequencies for high-fidelity audio equipment is 20 Hz, 50 Hz, 100 Hz, 400 Hz, 1,000 Hz, 1,500 Hz, 5,000 Hz, 7,500 Hz, 10,000 Hz, 15,000 Hz, and 20,000 Hz. These values will give you a good practical impression of the overall frequency response of the equipment you are testing. Remember, the human ear hears changes in frequency logarithmically, so the difference

between 100 Hz and 200 Hz sounds the same as the difference between 500 Hz and 1,000 Hz.

If you have to check the frequency response of an amplifier or
some other audio circuit but the output is designed for a load resistance other than the nominal 600 $\Omega$ assumed by the dB scale,
you can still perform a variation on this test, although it will be a
little more work, and you will have to do a bit more math. In this
case, you do not use the multimeter's dB scale. Use the regular ac
volts scale of the appropriate range. Measure the ac voltage across
the load resistor for each test frequency and note the measured
value. Also calculate the power value for each test frequency,
using the standard power formula:

$$P = \frac{E^2}{R}$$

where $P$ is the power in watts, $E$ is the voltage in volts, and $R$ is the
resistance in ohms.

It is usually helpful to first perform the test at a medium frequency, such as 1,000 Hz, and adjust the volume control for a convenient value, such as 4.0 V. Assuming an 8 $\Omega$ load resistance, this
corresponds to a power level of:

$$P = \frac{4^2}{8}$$
$$= \frac{4 \times 4}{8}$$
$$= \frac{16}{8}$$
$$= 2 \text{ W}$$

Table 8-1 shows a fairly typical set of results. Graph the output power and frequency, as illustrated in Fig. 8-6. The flatter this
line is, the better the frequency response of the amplifier. The amplifier used in this example is just fair, at best. Most modern audio
equipment will have a much better frequency response than this.
This rather poor-quality amplifier was chosen for this example
because it most clearly illustrates what we are talking about.
There will usually be the greatest attenuation at the lowest and
highest frequencies. There might be one or two dips or peaks in
the mid-range, as in our example.

Remember, the tone controls on any piece of audio equipment are designed specifically to change the frequency response
from a nominal flat line. For example, a bass tone control is used

**Table 8-1  Some typical results of the frequency response tests described in the text.**

| Test frequency | Measured output voltage | Output power |
|---|---|---|
| 20 | 2.5 | 0.78 |
| 50 | 2.9 | 1.05 |
| 100 | 3.4 | 1.44 |
| 400 | 4.2 | 2.20 |
| 1000 | 4.0 | 2.00 |
| 1500 | 4.0 | 2.00 |
| 5000 | 4.7 | 2.76 |
| 7500 | 3.8 | 1.80 |
| 10000 | 4.0 | 2.00 |
| 15000 | 2.8 | 0.98 |
| 20000 | 1.7 | 0.36 |

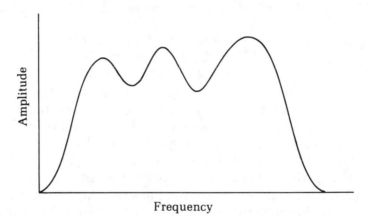

**Fig. 8-6**   *A graph is the most convenient way to indicate the frequency response of an amplifier, or other circuit.*

to selectively boost or attenuate the lowest frequency elements in the signal. To get accurate results, all such tone controls must be set to a flat position before performing the frequency response test.

# A quick test for the RF power output of a transmitter

An incandescent lamp of the proper wattage rating makes a good quick-and-dirty test for RF power output in high-power transmitters (anything from about 10 W up). Get a weatherproof lamp socket, a rubber-covered type with pigtail leads, and connect it to about 18 inches of coax with a PL-259 plug on the end. If you find a transmitter where there is doubt about the actual power output, screw this plug onto the antenna socket, put a lamp of the appropriate wattage in it, and key the transmitter. See Fig. 8-7.

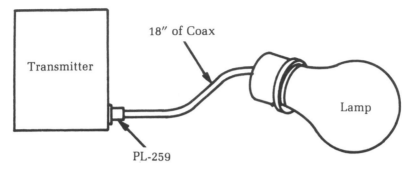

**Fig. 8-7** *The test hook-up for a simple check of RF power from a transmitter.*

If there is any RF power output, the lamp will light up. Since there is nothing but RF in the antenna circuit, a glowing lamp means that the transmitter has RF output and the trouble must be in the modulation, frequency, or something of that sort.

This light-bulb test is very useful for getting a quick start on those troubles where the problem could be in either the transmitter's final stage or the transmitting antenna. If the lamp lights up to a good white glow but there's very little output from the antenna, check the antenna and transmission line; that transmitter's okay. Although the mismatch between the transmitter and the lamp is theoretically something awful, you'll be surprised how little you have to change the tuning adjustments between the lamp and the regular antenna.

For CB transmitters, use a No. 47 pilot light. Even lower power? Try a No. 49.

# Measuring base bias voltage in high-resistance circuits

In a few audio-amplifier applications, you'll find very high value resistors used in the base voltage-divider circuits. One well-known make, for example, uses a 12 M resistor as part of the network. Even the 11 M resistance of a VTVM will upset this circuit if it is shunted across. Since it takes only a fraction of a volt to cause a big change in transistor currents, measuring the base bias directly is not a good procedure.

Set makers recommend measuring collector and emitter voltages as given in the service data. If these values are correct, then the base bias must be right. In other words, avoid getting into the very high resistance circuit to measure base voltage. Instead, measure the voltages that this voltage affects.

If you want to check out the base-bias circuit, remove the transistor and measure the large resistors for proper value, especially if the collector and emitter voltages are off-value.

# Finding breaks in multi-conductor cables

Multi-conductor cables are very handy in many applications. For example, a computer is connected to a printer via a multi-conductor cable, and a set of intercoms are interconnected with multi-conductor cables. A multi-conductor cable, of course, is nothing but a bundle of individually insulated conductors grouped together into a single cable and held together by an outer jacket of some sort, usually a hollow tube of the insulating material.

When interconnected devices aren't working together properly, the problem is often likely to be a break in a multi-conductor cable. How can a technician track down such a fault and identify which individual conductor in the multi-conductor cable is defective? A simple circuit for accomplishing this task is shown in Fig. 8-8. This circuit is so simple, there is no need for a separate

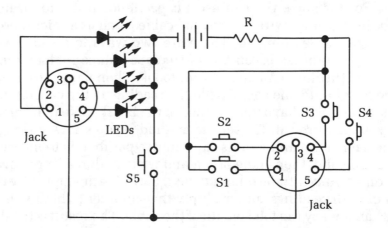

**Fig. 8-8** *This circuit can be used to locate a break in a multiconductor cable.*

parts list. Resistor R1 is simply a current-limiting resistor to protect the LEDs. Any value between 330 and 470 $\Omega$ would be suitable.

A standard 9 V transistor radio battery is used as the power source in this circuit. You could use a 6 V battery if you prefer, but 9 V batteries are small, inexpensive, and convenient. They are also very easy to find.

The two jacks are selected to match the plugs on either end of the cable to be tested. A five-conductor cable is assumed in this drawing, but you can easily increase or decrease that number to suit the specific multi-conductor cable you are working with. Except for the ground pin (or pins), each individual pin (conductor) gets its own LED on the left-hand side and its own push-button switch on the right-hand switch.

In operation, with the multi-conductor cable plugged into the two jacks, closing one of the switches should cause the appropriate LED to light up. If not, there is probably a break in that cable. If more than one LED lights up, there might be a short between those conductors.

The test for the ground connection is slightly different. Close switch S5. Nothing should happen. If any of the LEDs light up, there is a short in the cable, and it probably should be replaced.

# Finding a break in a coaxial cable with a capacitance tester

You can use almost any piece of test equipment for many applications. For instance, you can use a capacitance tester to locate a break in the inner wire of a coaxial cable, such as a microphone cable, coaxial lead-in, etc. There are two ways to do so. If you know the type of cable, you can find its capacitance per foot from a catalog or the *Radio Amateur's Handbook*. Then hang the capacitance tester onto one end. Divide the reading by the capacitance per foot, and you have the distance of the break, in feet, from the end you measured at. The second method, if you know the length of the cable but can't find its capacitance per foot, is to measure first at one end, then at the other, and write both readings down. Call one READING 1; the other READING 2. Divide READING 1 by the sum of both readings, and multiply that number (which will be less than one) by the total length of the cable. The result is the distance of the break from the end at which you took READING 1. Of course you can also use READING 2 in the same manner. This test is illustrated in Fig. 8-9.

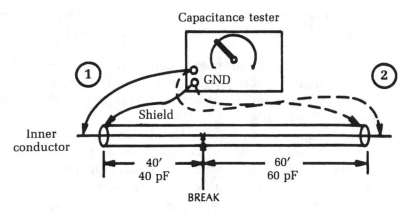

**Fig. 8-9**  *A capacitance tester can be used to find a break in a length of coaxial cable.*

Most breaks in microphone cables happen near one end or the other because of the bending and flexing near the plugs. To find out which end is broken and save tearing the plugs apart needlessly, take a capacitance reading from each end. One end will show a large capacitance, the other practically none; the latter is the broken end.

You also can use a capacitance test to find a broken wire inside the insulation of a two-conductor cable. Use the good wire as the "shield" and measure the capacitance of the broken wire to it. Otherwise, the method is the same.

## Measuring true rms voltages

Measuring dc voltages is generally straightforward. The only possible source of confusion is the reference point for the voltage measurement. However, the circuit's ground potential normally is used as the zero reference point and the few exceptions are usually clearly indicated.

Things get more complicated when you start dealing with ac voltages. An ac voltage, by definition, is constantly changing its value. There are several different ways to measure ac voltages, which are useful in certain circumstances. For instance, in some cases you are concerned with absolute maximum values. Here, you want to know the peak voltage reached during the ac waveform's cycle. For convenience, assume all waveforms are centered around 0 Vdc.

An ac voltage might be in the form of a sine wave with a peak voltage of 10 V. (It also would have a peak negative voltage of − 10 V.) To say you have 10 Vac would be misleading because the peak value is reached for only a tiny fraction of each cycle. For most of the cycle, the instantaneous voltage is considerably less than the peak value.

Because the voltage varies between +10 V and − 10 V, it covers a 20 V range. In some instances, it is useful to measure the signal as 20 V peak to peak. Once again, though, saying there's 20 Vac would be misleading.

The logical approach would be to take an average of the instantaneous voltages throughout a cycle. You can't use the entire cycle, however, because for a symmetrical waveform the positive half-cycle and the negative half-cycle will always cancel each other out, leaving an average value of 0 V.

For average ac voltages, only half of each cycle is considered. It has been mathematically proven that for a sine wave, the average voltage is always equal to 0.636 times the peak value. In our example of a sine wave with a peak voltage of 10 V, the average value would be 6.36 volts.

This is not unreasonable, and average values will allow you to meaningfully compare various ac voltages. Unfortunately, the formulas of Ohm's law do not hold true for average voltages. This is a major loss to the electronics technician because so much of circuit design and analysis is based on the relationships described by Ohm's law.

What you need in order to retain Ohm's law for ac voltages is a way to express ac voltage in terms that can be compared directly to an equivalent dc voltage. To find such an equivalent value, take the *root mean square* (rms) of the waveform. For a sine wave, the rms value is always equal to 0.707 times the peak value. In the example, the 10 V peak ac sine wave would be measured as 7.07 V rms. This signal would heat up a given resistor exactly the same amount as 7.07 Vdc through the same resistor. The voltage/current/resistance relationships defined by Ohm's law work out the same for rms values as for dc values. As a rule, ac voltage and current relationships usually are measured as rms values.

We have gone into some depth here because even experienced technicians occasionally get confused. For your convenience, the following is a handy comparison of the various ac measurements.

| | |
|---|---|
| rms | = 0.707 × Peak |
| rms | = 1.11 × Average |
| Average | = 0.9 × rms |
| Average | = 6.36 × Peak |

$$\text{Peak} = 1.41 \times \text{rms}$$
$$\text{Peak} = 1.57 \times \text{Average}$$
$$\text{Peak to peak} = 2 \times \text{Peak}$$

First, you have to convert an ac value form one from to another. Remember that these equations hold true *only* for sine waves. They cannot be used for other waveforms.

Most ac voltmeters are calibrated to measure ac voltages for sine waves. These meters will not be accurate for ac signals with any other waveshape. You are usually only concerned with making simple comparisons, so an ac voltmeter should suffice even for non-sine-wave signals. However, if you need a true rms value for a non-sine-wave signal, a standard ac voltmeter is virtually useless. Moreover, standard ac voltmeters are frequency dependent. They are tuned for accurate measurement at the standard line frequency (60 Hz).

These meters do the job fine for at least 75 percent of all service jobs. There are, however, occasions when they fall short. Until fairly recently, the technician didn't have much choice in the matter. If a standard ac voltmeter couldn't do the job, then you settled for compromised measurements, estimating, and making educated guesses.

Today, there are computer-driven meters that can perform the necessary mathematical calculations. Even better, the last few years have seen the development of specialized ICs that perform rms-to-dc conversion. An ac signal with almost any wave shape is fed into the input, and a proportionate dc voltage appears at the output. This dc voltage is the same as the true rms value of the input signal. These chips are starting to be used in multimeters, especially those of the digital variety. Two of the first rms-to-dc converter chips were the AD637, shown in Fig. 8-10, and the AD536AJD, which is illustrated in Fig. 8-11.

# Testing transistors

The vast majority of transistors are of the standard bipolar variety. *Bipolar* means that the semiconductor contains two pn junctions. The bipolar transistor is essentially a semiconductor "sandwich," as illustrated in Fig. 8-12.

Functionally, the bipolar transistor can be considered a pair of back-to-back diodes interconnected, as shown in Fig. 8-13. This is just an illustrative simplification. Normally, you cannot replace a transistor with two discrete diodes.

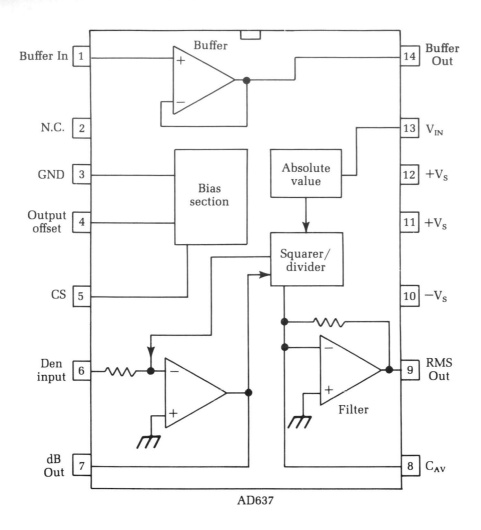

AD637

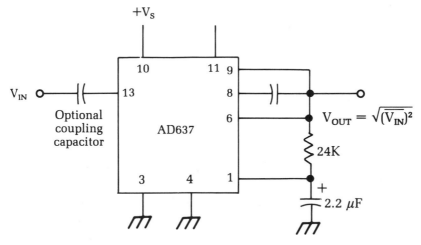

$$V_{OUT} = \sqrt{\overline{(V_{IN})^2}}$$

**Fig. 8-10**  The AD637 is a rms-to-dc converter IC.

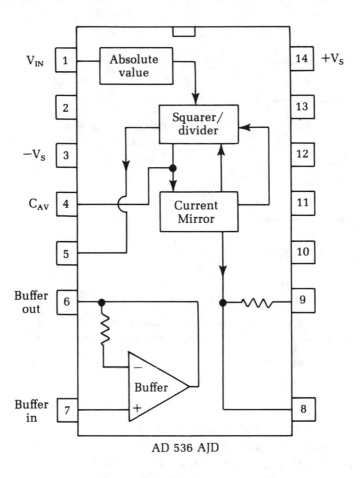

AD 536 AJD

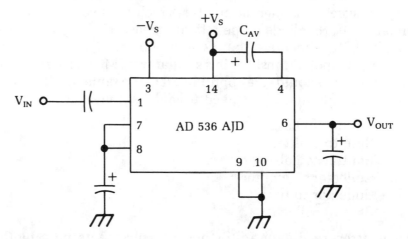

**Fig. 8-11**  *Another rms-to-dc converter IC is the AD536AJD.*

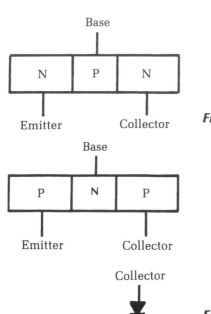

**Fig. 8-12**  *A bipolar transistor is essentially a semiconductor "sandwich."*

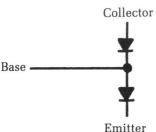

**Fig. 8-13**  *Functionally, the bipolar transistor is somewhat like a pair of back-to-back diodes.*

You can test a diode easily with an ohmmeter. Simply measure the resistance from one lead to the other, then reverse the leads and measure the resistance again. This process is illustrated in Fig. 8-14. You should measure a very high resistance in one direction and a low to moderate resistance in the opposite direction. If the resistance is high no matter which way you position the meter leads, the diode is open. If you measure a very low resistance in both directions, the diode is shorted.

Test a bipolar transistor in a similar way. Measuring between any two leads should give approximately the same results as with a single diode. In all, you need to make six measurements:

- Emitter to base
- Base to emitter
- Emitter to collector
- Collector to emitter
- Collector to base
- Base to collector

An incorrect reading on any of these measurements indicates that the transistor is bad and should be replaced.

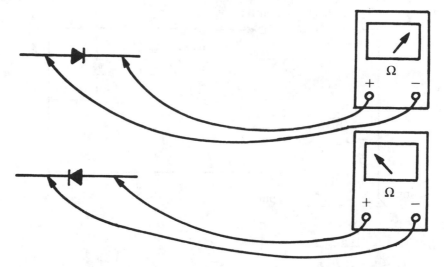

**Fig. 8-14**   *A diode or other single pn junction can be tested with an ohm-meter.*

In almost all cases, you should make these tests with the transistor out of the circuit. And other circuit resistances in parallel with the pn junction being measured will throw the reading off, especially if the parallel resistance has a low value.

The ohmmeter test works, but it is a little tedious to swap the test leads around six times. A dedicated transistor tester is a lot more convenient. Some very fancy transistor testers are available that measure transistor parameters such as alpha and beta.

These devices can be useful in certain cases, but for most repair work, you only need a simple "go/no-go" type of test. Many simple circuits have been designed for this purpose. Perhaps one of the simplest is the one shown in Fig. 8-15. You could easily build such a circuit in less than a half hour. The odds are good that you already have all the necessary components handy. The circuit is powered from an ordinary 9 V transistor radio battery.

Insert the transistor to be tested in the socket. Alternatively, you can replace the socket with test leads that have alligator clips on the ends. Attach the clips to the appropriate leads of the transistor. If the transistor to be tested is npn, set the dpdt switch to position A; set this switch to position B for pnp transistors.

Close the momentary-contact push button to test the transistor. If the transistor is good, the appropriate LED should light. If the LED does not light, assume the transistor is bad and replace it.

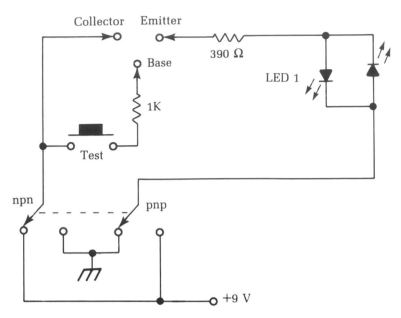

**Fig. 8-15**    *A circuit for performing simple "go/no-go" tests on bipolar transistors.*

You also can use this handy little circuit to identify a transistor's type. Simply test the transistor with the dpdt switch in both positions. In one position, an LED should light, while in the other position both LEDs should remain dark. You can identify the type of transistor by which LED lights. If neither LED lights at either position of the dpdt switch, the transistor is probably bad.

Both of these test procedures are for bipolar transistors only. If you try them on other transistor types, you will not get correct readings. Also, these tests work for out-of-circuit transistors, but you will not always get meaningful results if you attempt the test on a transistor wired into a circuit. This is a fairly serious limitation for servicing work.

Remember that all semiconductors are heat sensitive. If you're not careful, you can destroy a transistor when desoldering or resoldering. Always use a heatsink when soldering or desoldering around semiconductor components. The spring-loaded clip-on types, as shown in Fig. 8-16, are usually the most convenient to use. If you don't have one handy, fashion a make-shift heatsink out of a large paper clip. The larger the heatsink is, the more heat it will conduct away from the transistor. When in doubt, use a larger heatsink.

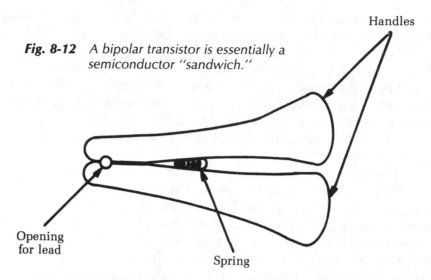

**Fig. 8-12**   *A bipolar transistor is essentially a semiconductor "sandwich."*

Handles

Opening
for lead

Spring

Obviously, it is impractical to desolder every transistor in a piece of equipment for out-of-circuit testing. Such a procedure should be limited to components where there is good reason to assume they're bad.

It usually isn't too difficult to isolate potentially troublesome transistors by making current and voltage tests. Standard values are usually printed in the schematic. Don't worry too much about minor errors. Look for voltage and current values that are significantly off. If a signal being fed into a transistor is correct and the voltage or current coming out isn't, then it is reasonable to suspect that transistor.

Don't be too quick to jump to conclusions, however. Often, a perfectly good transistor can look bad because of a problem in one or more of its associated components (resistors, capacitors, etc.). Even if you remove the transistor and find it tests bad, check the associated components anyway. Almost every electronics technician has had the frustrating experience of replacing a bad transistor only to have the new component soon blow out, too. Something else is wrong, causing the transistor to go bad.

Actually, such multiple problems are more common than you might suspect. Semiconductors are pretty reliable. Unlike tubes, transistors seldom go bad by themselves, although it has been known to happen occasionally. Usually something external has caused the transistor to go bad. In some cases, the transistor has been damaged by a transient that somehow managed to get into the circuit. In such an instance, there might be nothing else wrong with the circuit. When the transistor is replaced, the unit will work just fine. More frequently, however, the transistor's problem will have been caused by a resistor that has changed value, an open or leaky capacitor, or something similar.

Problems caused by other components might not always be obvious. Often it takes them a while to damage the new transistor. Run the equipment at least a half hour after making the repair to reduce the number of call-backs and irate customers you might have to face.

There is one final point about transistors you should be aware of. Unlike tubes, transistors are usually either definitely good or definitely bad. They are rarely weak. Once again, though, there are exceptions, so never be too hasty to jump to conclusions. On those rare occasions when you suspect a leaky transistor, you need a high-quality transistor tester to check it. A VOM/VTVM or the "go/no-go" tester like the one shown back in Fig. 8-15 won't be enough. A silicon transistor can have leakage as small as 10 to 15 $\mu$A. This small amount of current error is very difficult to measure by standard methods but can be sufficient to throw off proper circuit operation and cause all sorts of strange symptoms.

You should suspect a leaky transistor when everything in the circuit seems okay, but the circuit simply doesn't work the way it ought to. If all passive components are good, then an active device (such as a transistor) must logically be the culprit.

To perform a leakage test with a transistor, you almost always need to remove the component from the circuit. If you don't have a suitable transistor tester for the leakage test, but do have a duplicate of the questionable transistor, try substituting the duplicate in the circuit. If the circuit then works properly, it is safe to assume that the original transistor was leaky and should be discarded. On the other hand, if the symptoms persist, the odds are very strong that the problem lies elsewhere. It is extremely unlikely you'll encounter two leaky transistors (the original and the duplicate) in a row.

# Testing Zener diodes

There are many specialized semiconductor components. Often they require specialized testing procedures. In this section, we will look at methods for testing several popular semiconductor components that cannot be tested usefully in the same way as bipolar transistors and basic junction diodes.

The Zener diode is a special type of diode. Figure 8-17 shows a graph of the operation of an ordinary junction diode. If the applied voltage is near 0, the diode will not conduct. It will act like an open circuit, and no current will flow through it. When the applied voltage is increased past a specific positive voltage, the diode will start to conduct and will act pretty much like a short circuit. This forward-bias voltage is typically very small—about 0.3 V for germanium diodes and about 0.7 V for silicone diodes.

On the other hand, if you start at an applied voltage of 0 and increase the voltage in a negative direction, nothing will happen. The diode will be reverse-biased, and it will conduct little or no current. It continues to act like an open circuit. When some specific negative voltage is exceeded, however, the diode's junction will break down, and it will start to conduct very heavily, which will almost always permanently damage or destroy the diode. The reverse breakdown voltage varies with the construction of the specific diode in question, but it is always a relatively high value.

Figure 8-18 is a graph of the operation of a Zener diode. In the positive direction, the Zener diode works pretty much like an ordinary junction diode. Below about 0.6 V, the diode will not conduct, but when the applied voltage exceeds 0.6 V, the Zener diode's resistance will drop considerably, and a fairly large current will flow. So far, we have nothing special here.

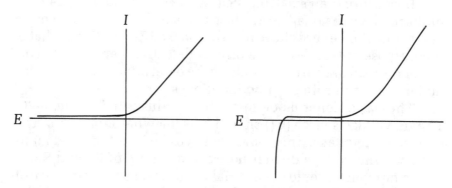

**Fig. 8-17** A graph of the operation of an ordinary semiconductor diode.

**Fig. 8-18** The Zener diode is a special type of semiconductor diode.

A Zener diode operates uniquely when it is reverse-biased. A small negative voltage won't have much effect. The high resistance of the reverse-biased diode means that virtually no current will flow through the device. When a specific key voltage ($V_z$) is exceeded, however, the junction resistance drops to a very low level, permitting the Zener diode to conduct fairly large currents, but the junction does not permanently break down. The Zener diode is not damaged by this negative conduction.

The most important feature is that this Zener voltage ($V_z$) does not change as the applied voltage is increased. For example, if a Zener diode is rated for 6.8 V, there will never be a reverse-biased voltage drop greater than 6.8 V across this component. Because of this unusual response, the Zener diode is used widely in voltage-regulator and reference-voltage applications.

Zener diodes are available with $V_z$ voltages ranging from about 2 to 200 V, and with power-handling capabilities of 0.25 up to 50 W. Even a Zener diode can be damaged if the reverse voltage is too large, but this usually takes a very large negative voltage. To prevent the Zener diode from self-destructing by drawing excessive current, you should use it with a current-limiting resistor.

To distinguish it from an ordinary junction diode, a special schematic symbol is used for the Zener diode. This symbol is shown in Fig. 8-19.

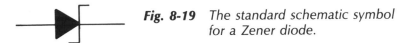

**Fig. 8-19**   *The standard schematic symbol for a Zener diode.*

Because of their small physical size, Zener diodes are usually unmarked. Occasionally, the electronics technician will need to determine the Zener voltage of an unmarked Zener diode. That is the purpose of this test procedure. A defective Zener diode will either be open or shorted, which can be determined with an ohmmeter, as with ordinary junction diodes.

The basic Zener diode test setup is illustrated in Fig. 8-20. The dc voltage source (B1) should be somewhat larger than the expected $V_z$ voltage. This source voltage can be attenuated manually via the voltage divider network made up of R1 and R2. It would be even better to use a variable dc power supply in place of B1, R1, and R2. Resistor R3's value should be selected to limit the current to no more than about 10 mA to 20 mA.

Notice that a voltmeter and a milliammeter are simultaneously used in this test. You can use two multimeters or you can use a multimeter and either a dedicated voltmeter or milliammeter. You must make the two measurements simultaneously.

Start out with a voltage of 0, or as close to 0 as possible. (Potentiometer R2 is adjusted for its maximum possible resistance.) You

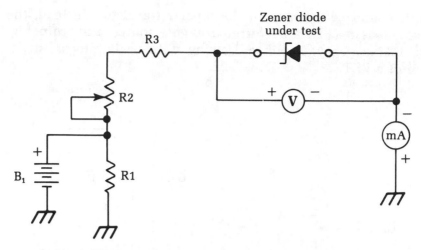

**Fig. 8-20**  *The basic test setup for checking Zener diodes.*

should read little or no voltage or current on the meters. Now, slowly increase the applied voltage (reduce the resistance of potentiometer R2). The reading on the voltmeter should increase with the increasing applied voltage, but the milliammeter should continue to give a 0 or near 0 current reading. Continue increasing the applied voltage. As this voltage nears the $V_z$ value, the current flow will start to increase. Once the $V_z$ value has been passed, the voltmeter should read $V_z$, and the milliammeter should indicate a drastic increase in the current flow. Increasing the applied voltage further should not change the reading of the voltmeter. The voltage the meter "got stuck on" is the $V_z$ value for the Zener diode you are testing.

If you don't get the results described, you did something wrong in the test procedure, or the Zener diode being tested is defective. Make sure the diode you are testing really is a Zener diode. Other types of diodes could give very strange results, if they are not damaged by the test procedure.

# Testing FETs

A field-effect transistor (FET) is quite different from an ordinary bipolar transistor. Its operating characteristics are much closer to those of old-fashioned vacuum tubes, offering the advantages of tubes without the excessive bulk, fragility, and heat that are inevitable with vacuum tubes.

The basic schematic symbol for a FET is shown in Fig. 8-21. Notice that, although it has three leads, they are not called *base*, emitter, and *collector* as in the bipolar transistor. Instead, the three leads of a FET are referred to as *gate*, *source*, and *drain*. Figure 8-22 is a very simplified drawing of the basic internal structure of a FET.

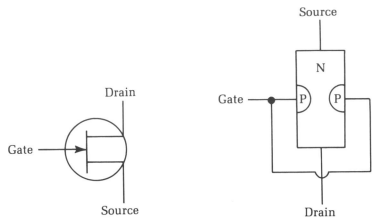

**Fig. 8-21**   *A FET is quite different from a bipolar transistor.*

**Fig. 8-22**   *A simplified drawing of the internal structure of a FET.*

The signal applied to the gate terminal of a FET controls the amount of electric current that can flow from the source to the drain. A negative voltage applied to the gate lead reverse-biases the pn junction, producing an electrostatic field (electrically charged region) within the n-type material of the component's body. This electrostatic field opposes the flow of electrons through the n-type section, acting somewhat like a partially closed mechanical valve in a plumbing system. The higher the negative voltage applied to the gate, the less current that is allowed to pass through the device from source to drain. Increasing the negative voltage "closes the valve" further.

This procedure is directly analogous to the action of a vacuum tube. The gate of the FET corresponds to the grid of the tube, controlling the amount of current flow. The source is the equivalent to the cathode (it acts as the source of the electron stream), and the drain serves essentially the same function as the plate (it drains off the electrons from the device). The path from the source to the drain is sometimes called the *channel*.

FETs have a very high input impedance, so they draw very little current. This is because of Ohm's law, which states:

$$I = \frac{E}{Z}$$

where $I$ is the current in amperes, $E$ is the voltage in volts, and $Z$ is the impedance (ac resistance) in ohms. Increasing the value of $Z$ will decrease the value of $I$.

The high input impedance of a FET means that it can be used in highly sensitive measuring and monitoring applications and in circuits where it is important to avoid loading down — drawing heavy currents from — previous circuit stages. This is another way in which FETs functionally resemble vacuum tubes.

The vast majority of FETs are of the n-channel type, as shown here, but FETs also can be made with a p-type channel and an n-type gate. These devices will work in the same way, except all polarities will be reversed. In this case, you must think of a positive voltage on the gate opposing the flow of holes through the channel. The schematic symbol for a p-type FET is the same as for an n-type unit, except the direction of the arrow on the gate lead is reversed.

The gain of a FET is usually identified as $Y_{fs}$, although this is not fully standardized, and other terms will be frequently encountered. The formula for FET gain is;

$$Y_{fs} = \frac{\Delta I_d}{\Delta V_{gs}}$$

where $I_d$ is the drain current, and $V_{gs}$ is the gate voltage for the FET in a common-source circuit. The small triangles ($\Delta$) represent the Greek letter delta and are used to indicate changing, rather than static, values.

A circuit for testing n-type FETs is shown in Fig. 8-23. To use this circuit for p-type FETs, you must reverse all polarities throughout the circuit. Notice that both a voltmeter and a milliammeter are used at once in this circuit. You can use two multimeters, or you can use a multimeter and either a dedicated voltmeter or milliammeter. You must make the two measurements simultaneously.

Set the dc voltage source for some specific voltage. The exact value is not too crucial, as long as it is within the FET's usable range. If you use a 9 V battery for B1, you shouldn't have to worry about this limitation. The voltmeter $V_{gs}$ shows the value, while the milliammeter indicates the $I_d$ value. You can find the static gain simply by dividing these two values:

$$Y_{fs} = \frac{I_d}{V_{gs}}$$

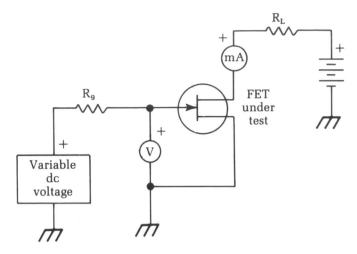

**Fig. 8-23**    *This circuit can be used to test n-type FETs.*

Unfortunately, this won't give you a very meaningful indication of how the FET will actually work in a practical circuit. So make a note of the measured $I_d$ and $V_{gs}$ values. Then change the applied voltage and take the measurements again. Determine the difference between the two $I_d$ values and the two $V_{gs}$ values, and use these change values to calculate the FET's dynamic gain. This test circuit is not intended for in-circuit tests of a FET.

# Testing SCRs

A silicon-controlled rectifier (SCR) is essentially an electrically controllable diode, or *rectifier*. In addition to the anode and cathode, there is a third lead, called a *gate*. The standard schematic symbols used to represent this semiconductor component are shown in Fig. 8-24. The circle around the symbol is optional and does not change the meaning of the symbol in any way.

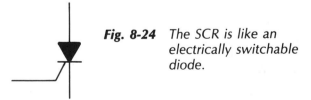

**Fig. 8-24**    *The SCR is like an electrically switchable diode.*

With no signal applied to the gate, no current can flow through the SCR from cathode to anode. A brief pulse, or trigger, signal of sufficient voltage applied to the gate terminal will turn the device on. The SCR will then conduct from cathode to anode, pretty much like an ordinary two-lead rectifier, even if the gate signal is totally removed. The only way to turn the SCR back off is to drop the voltage applied from cathode to anode below a specific threshold level, which is determined by the design of the SCR.

A rough equivalent circuit for a SCR made from a pair of bipolar transistors is shown in Fig. 8-25. This is a conceptual illustration only, and not a practical circuit. You cannot substitute a couple of bipolar transistors for an SCR in a working circuit.

Ordinary diodes are two-layer semiconductor devices. That is, there is a slab of p-type semiconductor and a slab of n-type semiconductor. A bipolar transistor is a three-layer device. In an npn transistor, two relatively thick slabs of n-type semiconductor are separated by a relatively thin layer of p-type semiconductor. In a pnp transistor, this arrangement is reversed — two relatively thick slabs of p-type semiconductor are separated by a relatively thin layer of n-type semiconductor.

An SCR is somewhat more complex in its internal construction. It is a four-layer semiconductor device, with two slabs of n-type material and two slabs of p-type material, interleaved as illustrated in Fig. 8-26. Because it is a four-layer semiconductor device, the SCR is one of a class of components known collectively as *thyristors*. Because the SCR's internal construction is so different from standard diodes or bipolar transistors, it is clear that you can't use the same test procedures for this type of component.

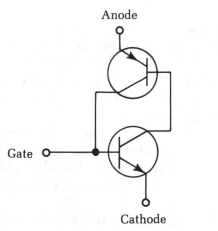

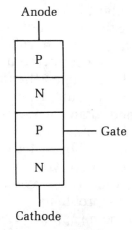

**Fig. 8-25**  *A rough (not functional) equivalent circuit for a SCR is comprised of a pair of bipolar transistors.*

**Fig. 8-26**  *A SCR is a four-layer semiconductor component, or a thyristor.*

It is fairly easy to test the gate turn-on voltage of an SCR. Just apply a variable dc voltage supply to the gate lead, as shown in Fig. 8-27. The load circuit illustrated is a simple LED indicator. When the SCR is turned on and conducting, the LED will light up. When the SCR is off, the LED will be off.

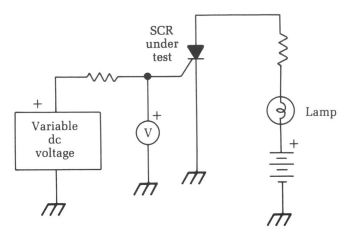

**Fig. 8-27**  *The circuit for testing the gate turn-on voltage for a SCR.*

Start the gate voltage at 0 and gradually increase it, while watching the LED. As soon as the LED lights up, stop increasing the voltage and look at the reading on the voltmeter. The voltmeter will directly indicate the voltage required to turn on this particular SCR.

Notice that once the LED starts glowing, increasing or decreasing the gate voltage, or even disconnecting the voltage source altogether, has no effect. To turn off the LED and reset the SCR, briefly open the NORMALLY CLOSED push-button switch in the LED circuit. This action will cut off the current flow from cathode to anode and turn off the SCR.

If the LED is always lit, or if it never lights, the odds are good that the SCR being tested is defective and should be replaced. Of course, before you draw any conclusions, you should double-check the wiring of your test circuit to make sure a mistake there isn't the problem.

In some applications, you will need to know the gate current ($I_G$) of a SCR. Because of the Ohm's law equations, the gate current will affect the required gate voltage to turn on the SCR. Matters can be somewhat complicated by the fact that the $I_G$ value is also dependent on the anode current, which is largely determined by the load circuit. All of this means that the gate voltage found in our last test is just a ball-park value. A somewhat (and sometimes significantly) different gate voltage might be required to turn on the SCR in a specific practical circuit.

Fortunately, it is not difficult to measure the gate current in virtually any SCR circuit. Just insert a milliammeter in series with the gate and gradually increase the applied current until the SCR just turns on. At this moment, the milliammeter will indicate the Igt value for that particular SCR in that particular circuit.

In manufacturer's specification sheets, the standardized value for the gate current is usually determined with a resistive load of 100 Ω and an anode voltage of 7 V.

Another important specification for a SCR is the forward-blocking voltage (Vdrm). This value determines the maximum voltage that can be applied across the SCR's anode and cathode. If this voltage is exceeded, the SCR will turn itself on even without a trigger signal at the gate.

In most practical SCR circuits, the Vdrm value should never be exceeded, since it essentially defeats the whole purpose of the SCR — that is, its gate controllability. If the forward voltage is too much above the Vdrm value, the component could be damaged. Usually, there is sufficient headroom between Vdrm and the absolute maximum voltage rating so that you won't have to worry too much about it, but you should be aware of the potential for damage.

You can use the circuit shown in Fig. 8-28 to determine the Vdrm value of a SCR. Notice that the gate terminal is essentially grounded, so the voltage on this terminal is always 0, by definition. There is no triggering signal fed to the SCR's gate. Slowly increase the applied dc voltage while watching the voltmeter carefully until the SCR fires and starts conducting. You must keep a

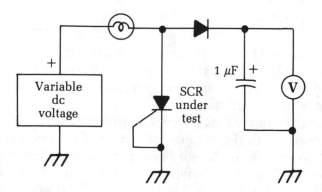

**Fig. 8-28**  *The circuit for testing the V$_{drm}$ value for a SCR.*

sharp eye on the meter during this test. Once the SCR starts conducting, the charge across the capacitor will start to leak off through the voltmeter's internal resistance, causing the reading to drift downward.

In practical SCR circuits, you also need to consider the reverse blocking voltage (Vrrm), the maximum voltage the SCR can withstand in its reverse-biased mode. It is roughly equivalent to the peak reverse voltage (PRV) specification for an ordinary diode. If the reverse-biased voltage is exceeded, the SCR will go into avalanche and will start conducting heavily. If this condition is allowed to continue, the component could self-destruct.

Figure 8-29 shows a circuit for testing a SCR's Vrrm value. Notice that the power supply for this circuit must be a constant current source to prevent the SCR from burning itself out during the test procedure. A test that destroys the component being tested obviously would not be of much practical use.

Increase the voltage slowly until the SCR starts to avalanche and starts to conduct the reverse-biased current. The voltage indicated on the voltmeter at that point is the SCR's Vrrm value.

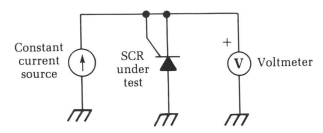

**Fig. 8-29**   The circuit for testing the $V_{rrm}$ value for a SCR.

# Testing ICs

Integrated circuits (ICs) can be tricky for the troubleshooter. Essentially, you have to treat an IC as a black box. All that matters are the external functions. The internal circuitry is irrelevant since there is no way to get at it.

Test the supply voltages and the inputs to the IC. If these voltages are correct but the outputs are wrong, it's a good bet the IC is bad and should be replaced. However, sometimes you can be fooled by loading. Occasionally, a problem in a later stage can throw off the reading at the ICs output pin(s). If there is any possibility of this, break the connection between the output pin in question and the later stages of the circuit. If the IC is okay, you should now get a correct reading at the output pin.

Similarly, a bad IC sometimes can affect the value at one or more of its inputs. Again, the way to find out for sure is to break the connection to the appropriate pin and measure the signal that would be presented to the ICs input if the connection was not broken. If this signal is now correct, the IC is bad.

With IC-based equipment, a schematic is even more important than for other types of servicing jobs. The more you know about how the IC is supposed to work, the better chance you have for properly identifying any problems. Your work area should include as many IC data books as possible. Virtually all IC manufacturers supply data sheets, and many offer extensive data books on their products.

If there is a particular type of IC you deal with frequently, you might want to rig up a simple dedicated tester for that particular IC. The easiest way is to construct a simple circuit around the IC, using a socket. Then, just plug the questionable IC into the socket and apply power. If the circuit works, the IC is good.

Like transistors, ICs are usually either good or bad. It is very unusual to find a weak IC, although it could conceivably happen. However, don't consider such a possibility unless everything else has been ruled out. The best approach then is to simply replace the questionable IC with a duplicate unit and hope for the best.

The test circuit should be as simple as possible. For example, if you work a lot with op amps, like the 741 and its descendants (which all have the same pin designations) you can put together a basic oscillator circuit with a small speaker as a tester. When you plug a good op amp IC into the socket and turn it on, a tone is produced by the speaker. This is a simple, clear-cut indication that things are working properly.

While on the subject of ICs, we should consider the question of sockets. Some experts swear by them, and others swear at them. IC sockets certainly make life easier on the service technician. ICs typically have 14 or 16 pins, and some have even more. Soldering (or resoldering) all of those tiny, closely spaced pins without creating solder bridges or overheating the delicate semiconductor crystal within the IC takes a sure, steady hand and fairly precise timing. It is also a tedious job, especially desoldering.

If a socket can be used, you can pop out an IC and replace it in less than a minute, without even plugging in the soldering iron. The use of a special IC extraction tool is highly recommended to avoid bending or damaging the pins. Some technicians routinely add a socket whenever they replace an IC. Then, if a similar repair is ever needed, the job will be considerably easier.

Unfortunately, IC sockets are not an unmixed blessing. The electrical contact between the pins and the traces of the pc board is not quite as good as with soldered components. In the vast majority of circuits, this probably won't make much difference. In some applications, however, especially those involving high frequencies, the use of a socket can lead to erratic operation. In portable equipment that can get knocked around, sockets are often undesirable. An IC can work its way out of its socket. It could get damaged. It could even short out some other part of the circuit, blowing out additional components. Finally, in many pieces of equipment, there just isn't enough room for a socket.

Use your own judgment on whether or not to add IC sockets. Some equipment already has sockets from the manufacturer. In this case, just be grateful for the relative ease of repair. There's no point in removing a socket installed by the manufacturer. (However, a socket added by another technician could conceivably be the source of problems in high-frequency circuits.)

You should exercise great care when replacing an IC, whether with a socket or not. Make sure the IC is correctly oriented. Applying power to a backward IC almost surely will result in disaster. Most ICs have a small notch indicating their "front" side, or they have a small dot over pin 1. Before removing an old IC, always make a note of its orientation.

When soldering an IC, be very careful. Don't apply too much heat. The pins are very closely spaced, so it is very easy to create an accidental solder bridge between two pins or adjacent pc traces. Never use too much solder and always heat-sink any IC before soldering it.

Make sure all of the pins go into the holes of the pc board or the socket. One might get bent up under the body of the IC. If this is not caught and corrected, there is a good chance the circuit will not operate correctly.

It pays to be extra careful when working with ICs. For one thing, many of them are very expensive, especially recently developed or highly specialized devices. Mass manufacture tends to bring down the cost of general-purpose devices. However, it's still smart to use extra caution even when replacing a "garden-variety" IC that only costs a quarter or fifty cents. Let's face it, replacing an integrated circuit is a royal pain. There's no reason to be careless to only end up having to do it over.

# 9
# Signal tracing and alignment tests

THE FIRST OUTPUT METER USED FOR ALIGNING TUBE RADIOS WAS AN AC voltmeter across the speaker-voice coil. A 400 Hz AM signal was used. This was nice, but noisy. Today, of course, you can always unhook the speaker and substitute an equivalent resistor. An easier way is to use a VTVM on the avc line. The avc voltage developed across the diode detector is always directly proportionate to the amount of signal. This is a negative-going voltage, and you'll find values from 1 V to 15 to 20 V.

Figure 9-1 shows a typical avc circuit as used in tube radios. This one is rather elaborate. You'll find fewer filter resistors and bypass capacitors in the smaller sets, but they all work the same way. The avc voltage appears at the top of the volume control, which is the diode load resistor. In most cases, you can pick up the avc at the mixer-grid section of the variable capacitor so you won't have to take the set out of the case. It can be picked up at any point along the avc circuit. Since the avc circuit has a very high impedance, you'll have to use a VTVM to get a readable deflection.

For best results, keep the input signal down to the point where it causes the smallest readable voltage on the avc; 1 V is a good average value. By doing so, you avoid the danger of overloading RF stages and flattening the response peaks.

## Checking the calibration of an RF signal generator

Do you want to set your RF signal generator exactly on 455 kHz to realign the IF stages of a radio? Do you have doubts as to the accu-

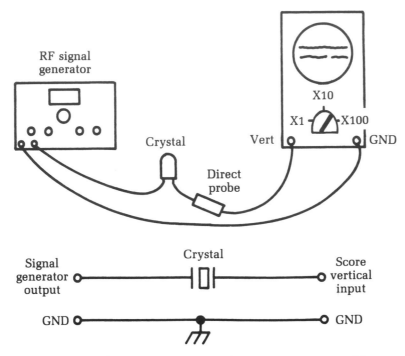

RF signal generator

X10

X1 · X100

Vert    GND

Crystal

Direct probe

Signal generator output

Crystal

Score vertical input

GND    GND

**Fig. 9-1**    *A method for setting an RF signal generator on a crystal frequency.*

racy of the generator's calibration? Then use the most convenient source of highly accurate test signals—broadcast stations. All AM radio stations are required by the FCC to keep their carrier frequencies within ±20 Hz of their assigned frequency. Most of them hold to within ±5 Hz.

To set your generator precisely on 455 kHz, get any radio that will pick up several stations. Its dial calibration doesn't matter; you are only going to use the receiver as an indicator. Choose a station as close to 910 kHz as you can find. Listen to it long enough to determine the call letters and then look up the carrier frequency of the station. Tune the signal generator to 455 kHz and then zero-beat this with the station carrier, using an unmodulated RF output. Your generator's second harmonic is beating with the station's fundamental.

If you can't find a station exactly on 910 kHz, locate one on each side as close as possible. Check each, note the error in the signal-generator dial. You can use this error to get the dial set on-frequency. For instance, if each station shows that the signal generator is one dial-marking low, then include this same amount of error when you set the generator for 455 kHz.

If you can't find stations close enough to the right frequency, try the third harmonic. At a 455 kHz fundamental, this is 1,365 kHz, so use a broadcast station at 1,360 kHz (all radio stations are on even number 10 kHz apart).

For high-frequency checking, use standard-frequency stations WWV or (in Hawaii) WWVH. You need a communications receiver that covers to 30 MHz, but the actual dial calibration of the radio is not important. These stations broadcast accurate test signals at 2.5, 5, 10, 15, 20, and 25 MHz. You can identify them easily by the 440 Hz beep tone, broken up by ticks at one-second intervals. Incidentally, these two stations also give standard time signals that are used the world over, in case you want to check your watch.

# Setting an RF signal generator on a crystal frequency

Every now and then you need to set a signal generator to an exact frequency, either for alignment or calibration purposes, or to check the calibration of the signal generator. If you have a crystal that operates at or near the frequency you need, setting the generator is easy. Some signal generators have provisions for plugging in crystals, but you can use a crystal even with a generator that does not have these provisions.

Connect the crystal between the RF output of the signal generator and the vertical input of a scope using a direct probe, as illustrated in Fig. 9-1. Set the RF output to maximum and turn the scope's vertical gain full up. Connect the ground leads of both instruments together, as shown. Now tune the signal generator very slowly back and forth over the frequency of the crystal. When you hit the exact frequency, you'll see the scope pattern increase in height.

The crystal is acting as a very sharp resonant filter. When you hit the right frequency, the RF voltage developed across it rises sharply; in typical tests, it might go from .05 to 0.5 V p-p. The scope need not be a wideband type; all you are looking for is an increase in pattern height. You'll have to tune the signal generator very slowly because the point of resonance is very sharp and you might pass it.

You can use an ac VTVM instead of the scope, or the DC-VOLTS range of a VTVM with a diode in series with the probe. The polarity of the voltage you read depends on how the diode is connected, but this isn't important; all you want is the peak.

You also can hook the crystal up in shunt: Connect the RF output lead and the scope input lead together, connect the ground leads together, and hook the crystal across them. Now you'll see a pattern on the scope (or a voltage on the meter) at all times. When you hit the crystal frequency, you'll see a very sharp dip in the pattern height or in the voltage reading.

When adjusted to the peak (crystal in series) or the dip (crystal in shunt), the signal generator is tuned to the frequency of the crystal and can be used as a standard for aligning RF or IF stages, receivers, or whatever is necessary. This test works pretty well with crystals up to 4.5 to 5 MHz, but above this frequency you usually run into low output from signal generators (the average signal generator uses harmonics in the VHF range, thus reducing the output by half or more). Therefore, there isn't enough RF power to get a usable reading on an indicator. The test would still work, with some kind of an amplifier or booster to get the output up to a readable level.

Although it isn't too reliable, this test can serve as an indicator of crystal output or activity (a broken or bad crystal won't respond at all). It also will roughly identify the frequency range of unknown crystals. If you test a crystal and get more than one reading, you can still identify the fundamental frequency of the crystal. It will make a deeper dip or give a higher output (in the series test) than any harmonic.

## Finding the exact point of a zero beat

When checking radio frequencies, you often want to find the exact frequency of a signal or set the bench signal generator exactly on a given frequency. To do so, check against a standard frequency of known accuracy. The easiest way is by zero-beating the unknown signal with the standard. All you need is a radio receiver that picks up the standard frequency. The standard itself can come from an accurate RF signal generator, from radio station WWV, etc. The receiver doesn't have to be accurately calibrated; it serves only as an indicator—a device for making the two frequencies beat against each other.

By ear alone, it's often hard to tell where the exact zero point of zero beat is. So hook an output meter (an ac voltmeter) to the radio output, as shown in Fig. 9-2. Next, couple the signal generator to be checked to the antenna of the radio, along with the standard. Sometimes just clipping its output lead to the insulation of the antenna lead is enough. Tune this signal to the test frequency,

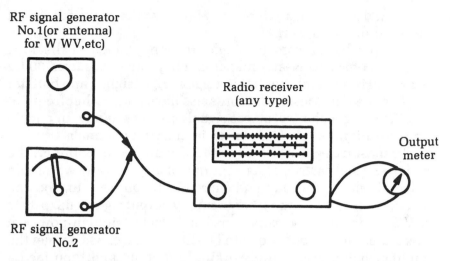

RF signal generator
No.1(or antenna)
for W WV,etc)

Radio receiver
(any type)

Output
meter

RF signal generator
No.2

**Fig. 9-2**  *To find the exact point of a zero beat, connect an ac voltmeter to the radio's output.*

listening for a zero beat in the receiver output as you tune. As the unknown signal approaches the frequency of the standard, you will hear a high-pitched tone that gradually goes lower as you get closer until you can't hear it at all. It then starts to go higher again as you tune past the standard.

Since the beat frequency goes down to zero, below the audible range, the exact zero point is hard to detect by ear. If you want to be very precise, tune for the lowest audible beat frequency and start watching the meter. When you reach a very low frequency, you'll see the meter needle start to wiggle as it tries to follow the beat note. At the same time, it will swing slower and slower. Tune for where the needle moves slowest.

# Test records: Good substitutes for the audio-signal generator

The average shop needs a high-quality audio-frequency signal generator, but seldom has one. It has to make-do with the 400 Hz audio output of an RF signal generator, although this will do the job for signal tracing and such work. However, you can get any kind of audio test signal you need at a very low cost, compared to the $300 or more for a high-quality AF generator. The source is a test record, and a great many different types are available — stereo, mono, or a combination of the two. With a suitable test record and an inexpensive record player, you're ready for almost any kind of audio work.

A typical test record has single-frequency bands for checking distortion or stylus wear, and a frequency-run series from 30 Hz up to 20 kHz. For stereo, it has left signal, right signal, and both for speaker phasing. Most test records also have many other frequencies. The single-channel stereo signals are very handy for checking separation, etc., and even for identifying channels.

In most cases, you won't even need an audio amplifier; the modern crystal cartridge has an output of up to 3 to 4 V, which can be fed directly into many audio circuits. If you want to, you can pick up a small, used amplifier to supply output signals up to 30 to 40 V for checking speakers and signal tracing in high-power af stages, etc. This amplifier needn't be hi-fi. You can use it to get the amplifier being tested into working condition, and then feed a signal directly from the cartridge into the input for distortion checks of the complete system with a scope.

# Using radio signals for testing hi-fi or PA systems

When testing hi-fi, stereo, or PA-system amplifiers, use a radio station as a signal source to let the amplifier "cook" for a while after repairing.

The input stage of the average high-gain audio amplifier acts as a detector for RF signals. Hook your TV antenna right across the input, as shown in Fig. 9-3 (you must have a closed circuit here). The lead-in and dipole make up a sort of long-loop antenna. Anything like a conical antenna, for example, which has no continuity across the lead-in, won't work; you'd get a tremendous buzz or hum.

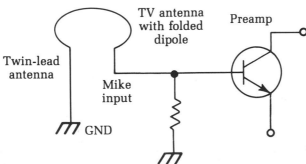

**Fig. 9-3**  *This circuit permits the use of radio signals for testing stereo or PA systems.*

This technique works best if there's a local radio station near enough to put a good signal into your shop area. If the signal is too weak, however, you can pick up music from a surprising distance by adding a detector diode in series with one side of the lead-in, or in shunt across the input.

If you have several strong locals, this system won't give a single clear sound. An alternative is to pick up the audio from the volume control of a small radio (either tube or transistor). You can set the gain wherever you want it and listen to the music as you cook the amplifier.

# Using a communications receiver as an RF signal locator or tracer

Every now and then you need an instrument that can find or trace an RF signal through a circuit, identify an unknown RF frequency, or signal-trace through a circuit to find where the trouble is located. There are special signal-tracing test instruments, but a standard communications receiver (of the kind we used to call a short-wave set) will do the job nicely.

Such receivers have antenna inputs that can be used on dipole antennas or convert to match 75 $\Omega$ coaxial cable; a shorting link on the antenna terminal board is used to make the conversion, as shown in Fig. 9-4. Get a piece of RG59/U coaxial cable 3 to 4 feet long and prepare one end to hook up to the terminal board.

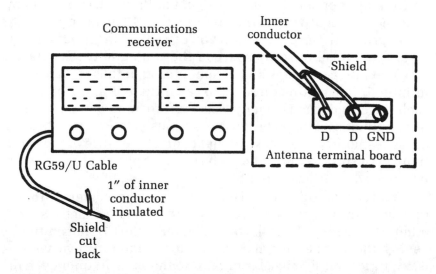

**Fig. 9-4** *A home-made probe for an RF signal locator or tracer.*

Cut off about 1 inch of the shield braid on the other end, exposing the insulated inner conductor. Don't take the insulation off; pull it out over the tip of the inner conductor so the wire can't make contact with anything.

Now, you can find and identify any RF signal within the tuning range of the receiver. For example, if you're checking a high-frequency oscillator circuit and want to know whether it's working on the right frequency — or working at all, for that matter — simply put the probe end of the cable close to the circuit and tune for the oscillator signal on the receiver dial. You can bend the end of the probe cable into a little hook that can be hooked over wires, etc., near the circuit under test.

An unmodulated RF signal sounds like a little thump in the speaker when tuned across. Tuned directly on it, you hear a rushing sound. A better way to identify it is to turn on the receiver's beat-frequency oscillator (bfo). Now if you cross an RF signal, you'll hear the characteristic beat note or sequel.

For an indication of the strength of the signal, hook a dc VTVM to the test receiver's avc circuit. For convenience, run a lead from the avc to a jack on the receiver panel or back apron, etc., and connect the VTVM there. If the receiver doesn't have avc, use the dc voltage developed across the audio detector; this is usually where the avc voltage comes from anyway. The voltage will go more negative as the input signal strength increases.

A receiver with a probe and an avc meter attached makes a handy little alignment indictor. For example, if you are tuning up an RF amplifier stage, feed in a signal at the desired frequency, tune the RF stage to it, and then pick up the RF signal output at the mixer grid with the probe. Now, you can adjust the input signal to a very low level to really sharpen the tuning of the RF stages.

You also can follow an RF signal through these stages to determine if it is getting through, to see if there is a loss or gain in each stage, and to make many tests that are impossible with other test equipment. The probe will not seriously detune any stage because it should never make actual contact. Because of the sensitivity of the test receiver, the probe picks up plenty of signal when it is placed near any circuit.

You can use this method for netting CB transceivers — that is, tuning all transmitters and receivers in a system exactly to each others' frequency. Tune the test receiver to the CB transmitter by keying the transmitter and tuning for maximum avc voltage. Next, tune your RF signal generator to the same frequency in the same way. Now feed the signal from the RF generator into the CB receiver, and tune it up for maximum output. You can use the test receiver as an output indicator by picking up the IF signal at the CB audio detector or you can use a scope or output meter. The CB transmitter itself can be used for this of course, but the RF signal generator will do a better job because its output can be controlled. Always make this kind of alignment adjustment on the smallest possible RF signal to avoid flattening the response curves or overloading the high-gain RF circuits.

In some VHF receivers, such as in two-way FM systems, frequency multiplication can be used to produce a VHF frequency from a low-frequency crystal oscillator. These receivers can be very confusing if you're not sure the multiplier coils are tuned to the correct harmonic! Set the dial of the test receiver to the correct harmonic frequency, hook the probe over the output lead of the multiplier coil, and tune for maximum.

## Using an ac voltmeter for gain checks and signal tracing

The ac voltmeter can be a handy instrument. The standard rectifier-type voltmeter has a good frequency range, making it especially useful for gain checks and signal tracing in audio amplifiers.

Feed an audio signal—about 400 Hz—into the input of an audio amplifier. Put a blocking capacitor of any size from 0.01 to 0.1 $\mu$F in series with the meter. (Rectifier-type meters are affected by dc; the capacitor blocks any dc and lets you read only ac or signal voltages.) Now you can start at either end of the amplifier and trace the signal through each stage to find out whether there is gain or loss.

This test is handy in amplifiers that are weak, but not dead. It's a rapid way to find the defective stage in a dead amplifier, too. Simply run through the circuit until you find the stage that has signal on the input but none on the output, and there you are.

This technique works with tubes or transistors and is especially useful in transistorized PC-board circuits. By using signal tracing first, you save the trouble of making voltage measurements and unsoldering parts before you have actually found the defective stage. The basic thing to check for is small signal on the input (base or grid) and a larger signal on the output (plate or collector), which would indicate that the stage does have gain. When you find a stage with no gain, or with a loss, that's the faulty one.

There is one exception to this rule in tube circuits and, to some extent, in transistors: the split-load phase inverter. As a rule, this stage is not designed to have any gain, but merely to divide and invert the signal. In the tube circuit, the load resistance is divided equally between the plate and cathode circuits. Thus, equal signal voltages appear on the two elements, but one signal is 180° out of phase with the other. There are several transistor circuits that do the same thing. To check for proper operation of such a circuit, measure the signal voltages on the input elements (grids or bases) of the following push-pull stage. The two signals should always be equal in amplitude. You can't check the phase without a scope, but chances are, if the signals are of equal amplitude, everything else is okay.

The ac voltmeter is also handy for finding troubles in stereo amplifiers. Feed the same audio signal into both inputs at the same time. Now measure the signal levels in corresponding stages of the two channels. At a given point, if you find one channel with a much lower signal than the other, that's the source of the trouble. Use the good channel as a guinea-pig to find out where the trouble is in the other channel.

The gain setting must be the same in both channels to prevent confusion. A good procedure is to turn both gain controls full on and then reduce the input signal level until the output level is about right. Don't overdrive; most stereo amplifiers have high gain and need only a small input signal. For the average phono input, 1 or 2 V is plenty; for a microphone input, much less will do —0.005 volt or less. Very high input signals can cause severe distortion or even damage transistors.

# Power output tests for PA and hi-fi amplifiers

Check the power output of PA amplifiers and hi-fis to see if that 30 W amplifier is actually able to deliver 30 W output. There is a simple test. After you have completed repairs and the amplifier is theoretically in first-class shape, hook up a properly matched load resistor across the output, feed in an audio signal, and read the power output by measuring the audio voltage across the load resistor. Ohm's law does the rest.

Figure 9-5 shows how the equipment is set up for this test procedure. You need a resistor that matches the output impedance of the amplifier and has a rating high enough to handle the

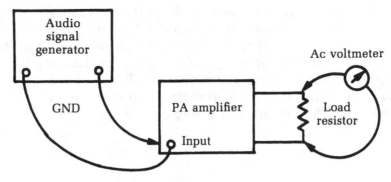

**Fig. 9-5**  *The setup for testing the power output of a stereo or PA amplifier.*

power, with a safety factor. For a 30 W amplifier, a 50 W resistor is good. You can get such resistors from surplus stores at reasonable prices. Otherwise, make them up from stock values. For instance, PA and hi-fi tube amplifiers usually have output transformers tapped at 4, 8, 16, and 500 Ω. Five 75 Ω 10 W resistors in parallel give 15 Ω at 50 W (for equal resistors, power ratings are totaled), and this is close enough for the 16 Ω tap. Five 2,500 Ω resistors in parallel give 500 Ω, etc.

With the load resistor hooked up, feed a low-level audio signal into the input; 1,000 Hz is a good frequency since most audio measurements are made at this frequency. Actually, the frequency doesn't make too much difference, as long as you're somewhere between 500 Hz and 5,000 or 6,000 Hz. (On most transistor amplifiers, especially the older ones, don't feed in a high-frequency signal — say, 15,000 Hz — at high power. The output transistors will overheat.) Remember this precaution: Never turn on an amplifier without the load resistor of the speaker hooked up. Even in tube amplifiers, you can burn out the output transformer in a very short time, and transistors can go in a fraction of a second if they are run without the proper load. Never short transistor outputs!

The service data gives the correct input level for many amplifiers. However, you are in the power output stage; can it deliver the rated power? To find out, hook an ac voltmeter across the load resistor and fire up the amplifier. If you're using a 15 Ω resistor and the amplifier is rated at 30 W, for example, you should read at least 21 V. Using $W = E^2/R$ and transposing to calculate $E$, you get $E^2 = 30 \times 15$, which yields about 21 V for $E$.

This is also a good voltage-amplification or sensitivity check. For example, on a phono input, you should get full power with the normal input level. With a high-output phono cartridge, this would be about a 2 V input signal. On a low-output microphone, it would be about 5 mV, etc. If you can get full output only by over-driving the input to two or three times normal, then one of the voltage amplifier stages isn't giving enough gain.

# 10
# Digital circuits

ELECTRONICS CIRCUITS CAN BE DIVIDED INTO TWO BROAD CATEGORIES: analog and digital. These two types are fundamentally different. In the past, virtually all electronics circuits were analog. Almost all of the circuits discussed so far in this book are analog circuits.

In the last decade or so, however, an electronics revolution has been taking place. Digital circuitry is becoming increasingly common, even in applications that were formerly solidly in the analog domain. Digital circuits are not used only in computers and calculators. They also show up in television sets, tape recorders, stereos, test equipment, alarm systems, and almost every conceivable type of electronics equipment.

Today's electronics technician has to be able to cope with digital circuitry. This is a frightening thought for many "traditional" technicians. Some technicians with long experience working with analog circuits feel intimidated by digital electronics.

There is really no need for such trepidation. Digital electronics is quite different from analog electronics in many ways, and servicing it often requires the technician to think in somewhat different ways. However, there is considerable overlap between digital and analog circuitry. More importantly, digital electronics is inherently simpler in concept than analog electronics. If you can understand the workings of a switch, you can understand digital circuits. Digital circuits are simply combinations of various switching functions.

In a digital circuit, a signal may take on one of only two possible values. A LOW signal is usually just slightly above ground potential, and a HIGH signal is a little bit below the supply voltage.

There are no other possibilities. These two states are given various names, but they all mean the same thing:

| Low | High |
|-----|------|
| 0 | 1 |
| No | Yes |
| False | True |
| Off | On |

(In some specialized applications LOW is called "1" and HIGH is called "0." This is done for convenience, and nothing is changed except the arbitrary names used.)

There is no ambiguity in a digital circuit. The signal is either clearly LOW or clearly HIGH. If it is not definitely one state or the other, then something is unquestionably wrong. An analog signal, on the other hand, is inherently ambiguous. Say a signal at a certain point in the circuit is supposed to be 5 V. Would 4.75 V be acceptable? How about 5.5 V? There is always a margin for error. In a digital circuit there is no margin for error. Only "yes" or "no" is permitted—"maybe's" can't exist.

Digital circuits are built around *semiconductors* called *gates* that are usually in IC form, so they can be treated as "black boxes." A digital gate produces a predictable output in response to the state of one or more inputs. The output/input pattern is frequently expressed in the form of a *truth table*, a notation of what the output should be for every possible combination of inputs.

The simplest digital gate is a buffer, analogous to the buffer amplifier found in analog circuits. A buffer amplifier has a gain of one, so the output is the same as the input. Similarly, a buffer gate does not change the state of the signal passing through it. The truth table for a buffer looks like this:

| Input | Output |
|-------|--------|
| 0 | 0 |
| 1 | 1 |

Note that there is no other possible input to this device, so there can be no other output condition. The truth table covers all possibilities. The schematic symbol for a buffer is shown in Fig. 10-1.

An inverter works something like the inverting input of an op amp. The output always has the opposite state as the input. The schematic symbol and truth table for an inverter are given in Fig. 10-2. The letter with the overscore is read "A not."

| Input | Output |
|-------|--------|
| 0 | 0 |
| 1 | 1 |

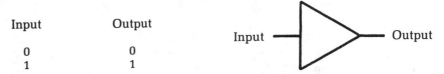

**Fig. 10-1**  *The simplest digital gate is the buffer.*

| Input | Output |
|-------|--------|
| A | $\overline{A}$ |
| 0 | 1 |
| 1 | 0 |

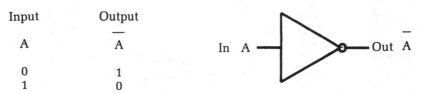

**Fig. 10-2**  *An inverter reverses the input state at the output.*

One-input gates are of limited use. More complex devices combine two or more inputs into one or more output(s). Multiple input gates might seem a little confusing at first, but they are easy enough to understand if you think about their names. For example, a two-input AND gate is shown in Fig. 10-3. The output is HIGH *if and only if* input A and input B are both HIGH. If either A or B (or both) is LOW, then the output must be LOW. This particle can be extended for any number of inputs. For example, Fig. 10-4 shows a four-input AND gate.

| Inputs | Output |
|--------|--------|
| A B | C |
| 0 0 | 0 |
| 0 1 | 0 |
| 1 0 | 0 |
| 1 1 | 1 |

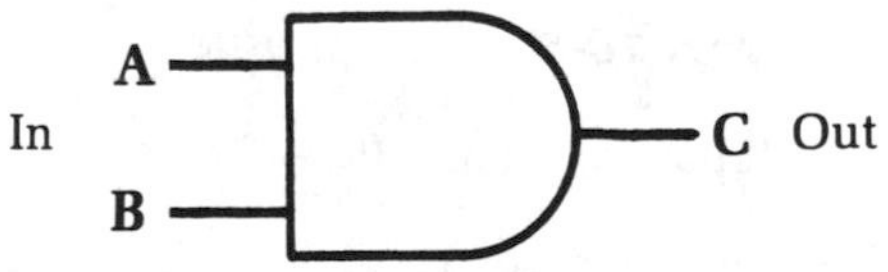

**Fig. 10-3**  *The AND gate.*

If you invert the output of an AND gate, as illustrated in Fig. 10-5, you get the opposite pattern. The output is HIGH *unless* both A and B are HIGH. This is called a NAND gate. The name is derived from "Not AND."

Another basic type of gate is the OR gate, shown in Fig. 10-6. The output is HIGH if either input A or input B is HIGH. The output of an OR gate can be inverted, as shown in Fig. 10-7, creating a NOR (Not OR) gate. The output is HIGH *if and only if* neither input A NOR input B is HIGH.

| Inputs | Output |
|--------|--------|
| A B C D | C |
| 0 0 0 0 | 0 |
| 0 0 0 1 | 0 |
| 0 0 1 0 | 0 |
| 0 0 1 1 | 0 |
| 0 1 0 0 | 0 |
| 0 1 0 1 | 0 |
| 0 1 1 0 | 0 |
| 0 1 1 1 | 0 |
| 1 0 0 0 | 0 |
| 1 0 0 1 | 0 |
| 1 0 1 0 | 0 |
| 1 0 1 1 | 0 |
| 1 1 0 0 | 0 |
| 1 1 0 1 | 0 |
| 1 1 1 1 | 1 |

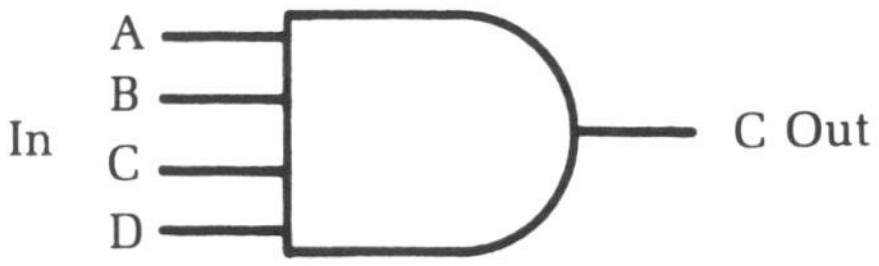

**Fig. 10-4** Digital gates can have more than two inputs.

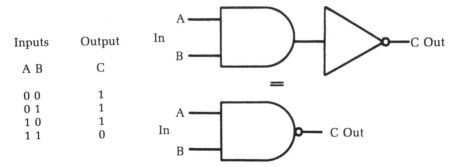

| Inputs | Output |
|--------|--------|
| A B | C |
| 0 0 | 1 |
| 0 1 | 1 |
| 1 0 | 1 |
| 1 1 | 0 |

**Fig. 10-5** The opposite of an AND gate is the NAND gate.

| Inputs | Output |
|--------|--------|
| A B | C |
| 0 0 | 0 |
| 0 1 | 1 |
| 1 0 | 1 |
| 1 1 | 1 |

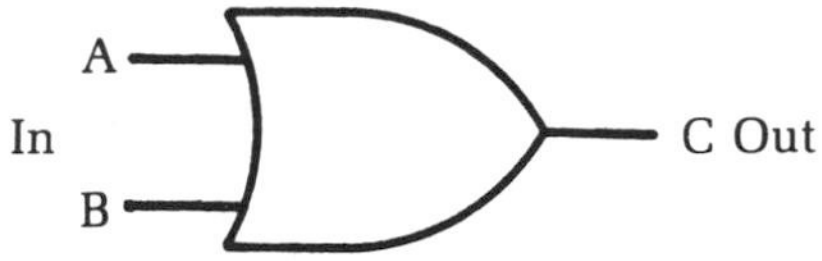

**Fig. 10-6** Another common type of digital gate is the OR gate.

A variation of the basic OR gate is the X-OR, or eXclusive OR gate, illustrated in Fig. 10-8. The output is HIGH if *either* input is HIGH, but not *both*. The X-OR gate can be referred to as a difference detector because the output goes HIGH only if the inputs are in different states. If the inputs are the same, the output will be low.

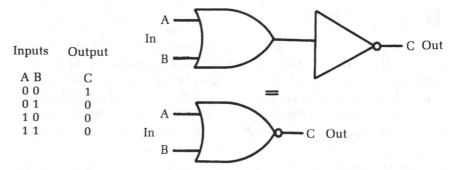

| Inputs | Output |
|--------|--------|
| A B    | C      |
| 0 0    | 1      |
| 0 1    | 0      |
| 1 0    | 0      |
| 1 1    | 0      |

**Fig. 10-7**  *Inverting the output of an OR gate results in a NOR gate.*

| Inputs | Output |
|--------|--------|
| A B    | C      |
| 0 0    | 0      |
| 0 1    | 1      |
| 1 0    | 1      |
| 1 1    | 0      |

**Fig. 10-8**  *The X-OR gate is a variation on the basic OR gate.*

ICs to perform all these basic gating functions are widely available and inexpensive Any combination can be created by combining various gates. Digital ICs that perform complex functions (such as counters, multiplexers, or even CPUs) contain multiple gate circuits internally. This discipline, *logic*, is the basis for computers.

Space does not allow a discussion of all the common digital ICs. In servicing digital equipment, you need to know what signal level should be at a given point under certain circumstances. Essentially this is the same idea used in testing analog circuits. There are just fewer possible signal values in a digital circuit.

In most digital circuits, you see a small capacitor across the power supply leads of each IC. This capacitor's job is to filter out noise and brief transients in the supply lines. Digital ICs are very sensitive to power-supply variations, even if they are very brief. If a piece of digital equipment starts behaving erratically, suspect problems in the supply voltage. Make sure the voltage is correct. Power supplies for digital circuitry should be very well regulated. Monitor the supply lines with an oscilloscope to see if there is excessive noise or transients. If just one IC seems to be misbehaving, the capacitor across its supply leads is a likely culprit. It could be open or leaky. Also make sure that there is a good connection to all of the IC's pins, especially if a socket is used.

# Logic probes

The simplest type of digital test equipment is the *logic probe,* a device that indicates the current logic state (HIGH or LOW) at a specific point in the circuit. A super simple logic probe made from a single inverter section is shown in Fig. 10-9. The LED lights when the probe detects a LOW signal. If the LED does not light, the logic state is assumed to be HIGH. A ground connection must be made between the probe and the circuit being tested. The supply voltage for the probe can be tapped off from the power supply of the circuit under test.

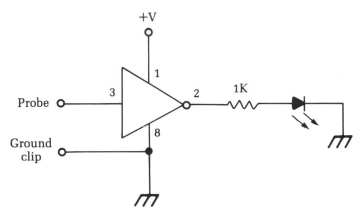

**Fig. 10-9** *A simple logic probe circuit. The pin numbers given are for a CD4049 hex inverter IC.*

Although functional, this simple logic probe leaves a lot to be desired. There is no way to distinguish between a no-signal condition and a LOW logic state. Also, in many digital circuits, the logic state changes back and forth at a high rate, often at frequencies above 1 MHz. When the LED is lit, it could indicate a HIGH state or a stream of rapid pulses.

An improved version of the circuit is illustrated in Fig. 10-10. This time there are two indicator LEDs. Together they can indicate four possible conditions:

| LED 1 | LED 2 | Indicated condition |
| --- | --- | --- |
| dark | dark | no signal |
| dark | lit | LOW |
| lit | dark | HIGH |
| lit | lit | pulses |

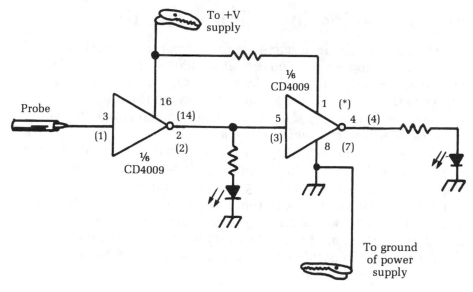

**Fig. 10-10**   *An improved version of the logic probe circuit shown in Fig. 10-9.*

# Monitoring brief digital signals

In many digital circuits, a signal might hold one state (HIGH or LOW) most of the time, but an occasional brief pulse to the opposite state will be crucial to correct circuit operation. Such pulses can be extremely brief—perhaps only a tiny fraction of a second. The LED in a simple logic probe such as those discussed in the preceding section would flash on and off too quickly to be visible to the human eye.

To detect such brief one-shot pulses, most commercial logic probes include a pulse stretcher circuit to extend the time the LED is lit. There are two basic approaches to pulse stretching. One is to use a timer in a monostable multivibrator mode. When the pulse is detected, it triggers the timer, which lights the LED for a fixed period of time, regardless of how brief the triggering pulse is.

The other approach uses a flip-flop. A flip-flop (or bistable multivibrator) can hold either logic state indefinitely. Each time it is triggered, the flip-flop reverses its output state. When the flip-flop detects a pulse, the LED lights and stays lit until a second pulse is received or the circuit is reset manually, usually via a small push button. The flip-flop is discussed in more detail later in this chapter.

# Signal tracing in a digital circuit

A logic probe is used in much the same manner as a signal tracer is used in an analog circuit. You can either start at the circuit's final output and work your way backward through the circuit, or you can start the input and work your way toward the output. In either case, when the signal is suddenly not what it should be, you have isolated the troublesome stage.

To perform signal tracing in a digital circuit, you need a very good understanding of what is supposed to be happening at each stage of the circuit. Work closely with the schematic and determine the function of each IC. If there is no signal, or if the output of a given IC remains constant regardless of the signals applied to any of its inputs, something is obviously wrong. No IC's function is to sit there and do nothing. Bad ICs will show one of these symptoms.

# Digital signal shape

All digital circuits work with square waves (or rectangular waves). Most digital gates cannot respond reliably to other waveshapes. If you have a digital circuit that is behaving erratically, monitor the digital signals with an oscilloscope. You should get a clean square wave without excessive ringing or noise. The tops and bottoms should be reasonably flat. The sides should be steep and straight. Distorted rise times or fall times can confuse many digital circuits.

# Power-supply problems in digital circuits

The majority of problems in digital circuits seems to be traceable to troubles in the power supply. Digital ICs tend to be very power-supply sensitive. Some types are more sensitive than others. TTL-type digital ICs are cheap and durable, but the supply voltage for these devices must be between 4.75 and 5.25 V. CMOS devices (which are becoming the norm) will accept a wider range of supply voltages, but they are still sensitive to severe noise or ripples. Transients (brief bursts) in the power line can cause erratic operation or even permanent damage to some of the digital chips.

Be very critical when taking measurements in a digital power supply circuit. If anything seems questionable, treat it as if it is bad. Better safe than sorry.

Again, to protect digital ICs from transients and power-supply noise, there is usually a small capacitor connected across the power supply pins of each individual IC. These capacitors are a fairly common source of trouble. If just one IC seems to be acting up, check the power bypass capacitor. It could be open or leaky.

In some inexpensive equipment, you cannot use bypass capacitors. If you want to guard against future problems, it might be worth your while to add the omitted capacitors yourself. They are certainly cheap enough and usually can be tack-soldered in place without too much difficulty. Be careful not to overheat the IC or you can destroy it. Always use a heat sink whenever you do any soldering near an IC. Mount the capacitor physically as close to the body of the IC as possible.

# Logic analyzers

A fairly new type of test instrument is the logic analyzer. This instrument can be thought of as the digital equivalent of an oscilloscope. It permits the technician to view the signals in a digital circuit directly.

In a nutshell, what the logic analyzer does is to sample digital signals and store them for later review. The stored signals are in the form of 1's and 0's.

A logic analyzer can be used in any digital circuit. It is most useful in microprocessor-based circuits. Generally the logic analyzer is used to monitor the signals appearing on the address or data buses. Several signals are monitored simultaneously. Some logic analyzers display the stored data directly on an internal display. Others feed their output data to an ordinary oscilloscope for display.

The logic analyzer is really too complex to discuss in full detail here. It would take a book to completely cover its operation and use. If you do much work with digital (especially microprocessor-controlled) equipment, you should look into acquiring a logic analyzer.

# Pulse generators

Pulse generators are becoming increasingly common on test benches. Basically, a pulse generator is a special-purpose signal generator. The circuitry is not dissimilar to the function generator.

A pulse generator puts out clean square waves and rectangular waves with steep rise times and fall times. The signals are suitable for use as test signals in test equipment, and a number of triggering modes are usually offered for various types of tests. A pulse generator sometimes can be used as a regular analog function generator, also.

Most of the better pulse generators offer a number of special features. One feature that can be useful for checking an erratic digital circuit is variable rise/fall time. Starting from the minimum settings (steepest sides on the waveform), gradually increase the times until the device being tested starts showing the faulty symptoms. Now you know how much the signal can be distorted without trouble. Then use your scope to find the stage that distorts the pulse past the allowable amount.

A full discussion of the pulse generator is beyond the scope of this book, but if you do much work with digital circuits, a good pulse generator on your workbench is a handy tool.

# Flip-flops

Ultimately, all digital circuits break down into gates, but for most practical applications, dealing with individual gate ICs would be unwieldy at best. In sophisticated applications, the sheer bulk involved would render the circuit utterly impossible to build and operate under any realistic circumstances. For this reason, multiple gate units are combined into single ICs. A central processing unit (CPU) for a computer might contain the equivalent to thousands or even millions of gates within a single 40-pin IC.

Of course, there are many digital devices between the simple gate IC and the sophisticated CPU. Devices at the lower end of the scale generally have wider applications because they are less inherently specialized.

One of the most important of these medium-level digital devices is the flip-flop. More technically, the flip-flop is a bistable multivibrator. A multivibrator has two possible output states: HIGH and LOW. The output signal is never between these two extremes.

A monostable multivibrator has one stable state. It holds this state indefinitely. When the monostable multivibrator is triggered by an external signal, the output jumps to the opposite, unstable state for a fixed period of time, determined by specific component values within the multivibrator circuit.

An astable multivibrator has no stable states. Its output con-stinuously switches back and forth between the HIGH and LOW states at a regular rate. The timing of these switchovers is deter-mined by specific component values within the multivibrator circuit. In most cases, no external input signal is applied to an astable multivibrator.

A bistable multivibrator has two stable output states. This type of circuit can hold either a LOW output or a HIGH output indef-initely, as long as power is continuously applied to the circuit. Each time the bistable multivibrator circuit receives a trigger pulse (external input signal), the output reverses its state, from LOW to HIGH or from HIGH to LOW. This new output state will be held indefinitely until a new trigger pulse is received to reverse the output state once more.

In effect, the bistable multivibrator, or flip-flop, "remem-bers" the last output state to which it was set. This type of circuit is therefore sometimes referred to as a *one-bit memory*. Another common name for this device is a *latch*. In practical usage, the rather casual sounding *flip-flop* is the most frequently employed term for this type of circuit.

Some electronics technicians can understand digital gates fairly easily but find flip-flops a bit confusing, especially there are several different types of flip-flops that function in significantly different ways, despite their major similarities. A technician ob-viously can't trouble-shoot and service a flip-flop circuit if he or she doesn't know how it is supposed to work in the first place. For this reason, it seems reasonable to briefly discuss the principles of flip-flops and the various types of this device used in modern electronics circuits.

## RS flip-flop

Figure 10-11 shows one of the simplest flip-flop circuits. It is made up of a pair of two-input NAND gates. This is a RS flip-flop. Notice that it has two inputs, labeled "S," for Set, and "R," for Reset. The circuit also has two outputs labelled and Q and $\overline{Q}$, or NOT Q. (The bar over a logic term always indicates inversion or negation.) The Q and $\overline{Q}$ outputs are always at opposite states; that is, if Q is HIGH then $\overline{Q}$ is LOW, and vice versa.

In examining the operation of this RS flip-flop circuit, let's assume that when power is first applied to the circuit both the S and R inputs are HIGH and output Q is LOW. Of course, this means that output $\overline{Q}$ must be HIGH.

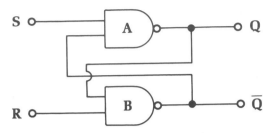

**Fig. 10-11**   *A simple RS flip-flop circuit.*

Notice that the inputs to NAND gate A are S (HIGH) and $\overline{Q}$ (HIGH). This means that the output of this gate (Q) will remain LOW, until the input conditions are changed. Similarly, the inputs to NAND gate B are R (HIGH) and Q (LOW), so its output ($\overline{Q}$) remains HIGH. The circuit will be latched in this state as long as both inputs (S and R) are held HIGH.

If input S is now brought LOW, the situation changes. The output of gate A (Q) is changed to HIGH, which, in turn, forces the output of gate B ($\overline{Q}$) to go LOW.

Even if input S is changed back to the HIGH state, the flip-flop circuit's output states will not change because of the way they are fed back to the inputs of the gates. The circuit is latched into this new state, and any changes in the signal at the S input will have no effect on the output signals.

Next, let's change input R to LOW. This will change the output of gates B ($\overline{Q}$) back to HIGH, and gate A's output (Q) will go LOW once more. Further changes in the R input will have no effect on the circuit's outputs.

In other words, a 0 (LOW) input at S sets the flip-flop (Q = HIGH and $\overline{Q}$ = LOW), while a 0 input at R resets the flip-flop (Q = LOW and $\overline{Q}$ = HIGH). A 1 (HIGH) at both inputs will latch the circuit in its present state. The outputs will not change states.

Notice that a 0 at both inputs is a disallowed state. The output will be unpredictable under this condition. If this situation inadvertantly occurs because of some sort of circuit defect in an earlier stage of the digital circuit, all sorts of strange things might happen, or the circuit might simply become unreliable—sometimes it works and sometimes it doesn't.

As with digital gates, the operation of a flip-flop can be summarized in the form of a truth table. Following is the truth table for the simple RS flip-flop circuit described previously:

| Inputs | | Outputs | |
|---|---|---|---|
| R | S | Q | $\overline{Q}$ |
| 0 | 0 | Disallowed state | |
| 0 | 1 | 0 | 1 |
| 1 | 0 | 1 | 0 |
| 1 | 1 | Latch—No change | |

In some technical literature, the RS flip-flop might be referred to as a *set-reset flip-flop*. These terms are interchangable.

## JK flip-flop

Another common type of bistable multivibrator used in digital circuits is the JK flip-flop. The "JK" is used to distinguish this type of flip-flop from the RS type. J and K apparently don't stand for anything in particular.

The JK flip-flop is somewhat more complex than the RS flip-flop. It has five inputs and two outputs, as illustrated in Fig. 10-12. As in the RS flip-flop, the outputs of the JK flip-flop are labeled "Q" and "$\overline{Q}$." Once again, $\overline{Q}$ is always the complement of Q. The five inputs to a JK flip-flop are: preset, preclear, J, K, clock.

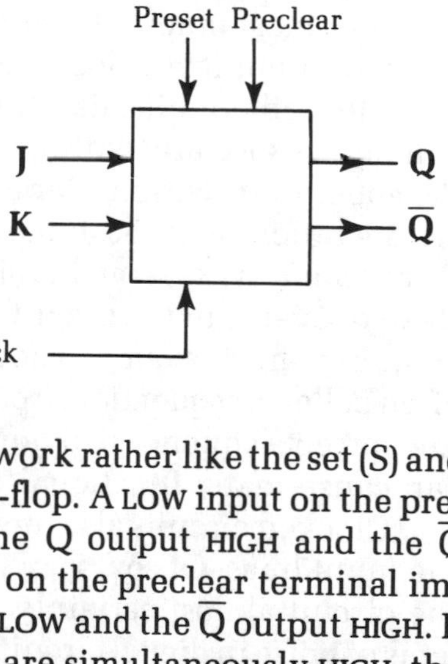

**Fig. 10-12**  *The JK flip-flop is somewhat more sophisticated than the RS flip-flop.*

The preset and preclear inputs work rather like the set (S) and clear (or reset, R) inputs on a RS flip-flop. A LOW input on the preset terminal immediately forces the Q output HIGH and the $\overline{Q}$ output LOW. Similarly, a LOW input on the preclear terminal immediately forces the Q output to go LOW and the $\overline{Q}$ output HIGH. If both the preset and preclear inputs are simultaneously HIGH, the output states will be determined by the other three inputs.

Putting both the preset input and the preclear input in a LOW state simultaneously is a disallowed state. The circuit will become confused, and the output states will be unpredictable, resulting in very erratic operation of the digital circuit at best.

The preset and preclear inputs can be activated at any time. They are not affected by the clock input (discussed in the next paragraph). These inputs take precedence over the other three inputs, which are ignored by the flip-flop circuit unless both preset and preclear are HIGH.

The J and K inputs are *clocked inputs*, which means they can have no effect on the outputs until the clock input receives the appropriate pulse signal. When the clock input is triggered, the flip-flop circuit looks at the J and K inputs and responds accordingly.

There are two basic types of clocking: level clocking and edge clocking. In a level-clocking system, the clock input is triggered by the actual logic state of the input signal. It might be designed to trigger on either a HIGH state or a LOW state, but not both. The clock will remain activated for as long as the input is held at the appropriate logic level.

Edge clocking, on the other hand, is triggered by the transition from one state to the other. Either the LOW to HIGH (positive edge) transition or the HIGH to LOW (negative edge) transition can be used (but not both, depending on the design of the specific circuit within the IC itself. Obviously, an edge-clocked device is activated for a much shorter time period than a level-clocked device. For most practical digital circuits, edge clocking is far more common than level clocking.

Clocked circuits, like the JK flip-flop, have a number of advantages, especially within large systems. First, by triggering all the subcircuits within a large system from the same clock signal, all operations can be forced to stay in step with each other throughout the system, helping to prevent erroneous signals. Instead of events in the circuit just happening whenever they happen, they must occur at a specific time. Such a clocked circuit is often called a *sequential circuit* because the signals must occur in the correct sequence. Obviously, a flaw in the clocking circuitry can cause many functional defects in the system as a whole.

Clocked circuits also tend to be a bit less sensitive to noise on the input lines. Of course, if noise occurs on the clock signal line, the circuit almost certainly won't function properly. The problems of noise in digital circuits will be discussed later in this chapter. The following is the truth table for a JK flip-flop

| Inputs | | | | | Outputs | |
|---|---|---|---|---|---|---|
| ps | pc | C | J | K | Q | $\overline{Q}$ |
| 0 | 0 | x | x | x | Disallowed state | |
| 0 | 1 | x | x | x | 1 | 0 |

| 1 | 0 | x | x | x | 0 | 1 |
| 1 | 1 | N | x | x | No change | |
| 1 | 1 | T | 0 | 0 | No change | |
| 1 | 1 | T | 0 | 1 | 0 | 1 |
| 1 | 1 | T | 1 | 0 | 1 | 0 |
| 1 | 1 | T | 1 | 1 | Reverse states | |

In this truth table, a *0* indicates a logic LOW signal, and a *1* indicates a logic HIGH signal. An x means "Don't Care." The state of a "Don't Care" input has no effect on the output signal. The input signal marked x may be either a LOW or a HIGH — it doesn't matter. N means the clock is not triggered, and T means the clock is triggered. C stands for the clock input, ps is the preset input, and pc is the preclear input.

Notice that if both the J and K inputs are HIGH (assuming the preset and preclear inputs are both HIGH, of course), the outputs will reverse states each time the clock is triggered. For example, let's say the Q output starts out as a 0 ($\overline{Q} = 1$). On the first clocking pulse, Q will become a 1 ($\overline{Q} = 0$). The second clocking pulse will change Q back to a 0 ($\overline{Q} = 1$). The third clocking pulse changes Q to a logic 1 again ($\overline{Q} = 0$).

## D-type flip-flop

The JK flip-flop is clearly quite useful and versatile. In most practical applications, the preset and preclear inputs aren't used (they are hard-wired to HIGH), and there is no disallowed state to potentially confuse matters with the clocked inputs. However, the JK flip-flop's requirement for two inputs (J and K), in addition to the clock signal, sometimes can be inconvenient in certain applications.

The problem can be solved by using yet another type of flip-flop circuit, known as the D-type flip-flop. The D stands for data. Figure 10-13 illustrates how a D-type flip-flop can be made from a JK flip-flop and an inverter. Notice that, in this circuit, the J and K inputs must always be at opposite states. If J is HIGH, then K must be LOW, and vice versa.

Following is the truth table for a D type flip-flop:

| Inputs | | Outputs | |
| --- | --- | --- | --- |
| C | D | Q | $\overline{Q}$ |
| N | x | No change | |
| T | 0 | 0 | 1 |
| T | 1 | 1 | 0 |

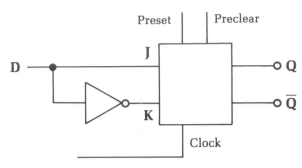

**Fig. 10-13**   *A D-type flip-flop can be made from a JK flip-flop and an inverter.*

In this truth table, *C* is the clock input. *N* means the clock is not triggered, and *T* means the clock is triggered. The *x* stands for "don't care." The D input is ignored by the circuit except when the clock input is properly triggered. Most D-type flip-flops also have preset and preclear inputs that function exactly as in a JK flip-flop, discussed previously.

As you can see from the truth table, a D-type flip-flop is fairly simple, but it can be made even simpler. You don't always need an external D input signal. A D-type flip-flop can operate on just the clock signal, using the configuration shown in Fig. 10-14. The D input is fed by the $\overline{Q}$ output. The clock must be edge-triggered in this type of circuit, so there is only time for a single-state change during each clock pulse.

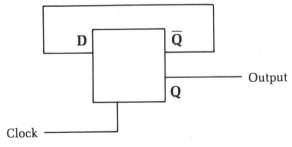

**Fig. 10-14**   *A D-type flip-flop can operate on just the clock signal, dividing the clock frequency by two.*

Assume that the Q output starts out HIGH; therefore, the $\overline{Q}$ output must be LOW. This signal is fed back to the D input, forcing it LOW, too. Nothing happens until the clock is triggered. Then the flip-flop circuit looks at the data on the D input. Since this a 0, the truth table tells us the Q output must go LOW, and $\overline{Q}$ becomes HIGH.

This HIGH signal is fed back to the D input, but by this time the clocking pulse is gone, so the flip-flop waits until the next trigger signal is detected When the clock is triggered a second time, the HIGH signal on the D input changes Q back to HIGH, and $\overline{Q}$ drops back down to the LOW state, and we're back to where we started. This pattern will repeat indefinitely, as long as power is applied to the circuit and the clock trigger pulses keep coming.

The input and output signals for this circuit are illustrated in Fig. 10-15. Notice that both signals are in the form of square waves. Actually, the input might be any rectangular or pulse wave, but assuming the clock signal has a regular frequency, the output signal will always be a true square wave.

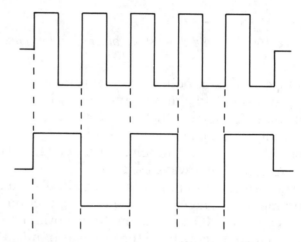

**Fig. 10-15**   *Some typical input and output signals for the circuit of Fig. 10-14.*

Notice that it takes two complete input (clock) cycles to produce one complete output cycle. This means the output frequency is exactly one-half the input frequency. For this reason, this type of circuit is often called a frequency divider; it divides the clock frequency by two.

This simple circuit is also a digital counter, albeit of a very trivial sort. It can count only two steps before it resets itself and starts the count over. By itself, it is of limited value as a counter, but you can cascade multiple D-type flip-flops in series to create more advanced binary counters. (In the binary numbering system there are just two numerals: 0 and 1.)

A three-stage binary counter made from D-type flip-flops is shown in Fig. 10-16. The preset and preclear inputs of each of

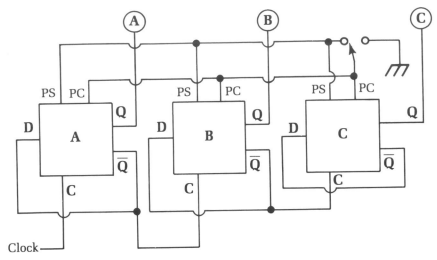

**Fig. 10-16**  *A simple three-stage binary counter circuit.*

three flip-flops in this circuit are tied together, and the preset inputs are connected to the supply-voltage line, so they are always held HIGH. Remember that each IC package also must have its power-supply pins connected to the power supply, even if this is not explicitly shown in the schematic diagram. The power-supply connections are always assumed.

The three preclear inputs go to a single SPDT switch, preferably with momentary contacts for one NC connection (shown in the diagram) and one NO connection. In its normal position, the switch sends a logic 1 (HIGH) to all the preclear inputs. When both the preclear and preset inputs of a D-type flip-flop are HIGH, the output states are determined by the D input and the clock input. When the switch is moved briefly to its other (NO) position, however, a logic 0 (LOW) is applied to the preclear inputs, forcing the Q output of each flip-flop stage LOW (and each $\overline{Q}$ output goes HIGH), regardless of their previous states. This switch clears the counter by forcing it to a 000 output state.

As you look at the operation of this simple counter circuit, you will start out with this initial 000 condition, and see what happens on each successive clock pulse. Assume that the clock inputs trigger on the positive edge (that is, on the LOW to HIGH transition).

**Clock pulse #1**  Flip-flop A has a 1 on its D input (since D is tied to $\overline{Q}$ in each stage). When the first clock pulse is received, QA goes to 1 and $\overline{Q}$ goes to 0. Flip-flop B is clocked by the 0 to 1 transition of flip-flop A's $\overline{Q}$ output. Since $\overline{QA}$ just went from 1 to 0, flip-flop B is not triggered, and its output state remains the same as before. Similarly, flip-flop C is clocked by $\overline{QB}$. This signal has not changed during this clock pulse, so flip-flop C's output is also unchanged. Reading the outputs from right to left (i.e., C — B — A), we have a binary count of 001, or one.

**Clock Pulse #2**    $\overline{QA}$ is a 0. This is fed back to DA. When the triggering clock pulse is received, QA goes to 0 and $\overline{QA}$ goes to 1. Since QA has made a 0 to 1 transition, flip-flop B is also triggered. The 1 on the DB input (from $\overline{QB}$) changes QB to a 1 and $\overline{QB}$ to a 0. Flip-flop C is not triggered. Reading the outputs from right to left, we now have a binary count of 010, or two.

**Clock Pulse #3**    This pulse behaves like Clock Pulse #1, changing QA to a 1 and $\overline{QA}$ to a 0 and leaving flip-flops B and C unchanged. The outputs now read 011, or three.

**Clock Pulse #4**    DA is being fed a 0 (from $\overline{QA}$), so QA changes to a 0 and $\overline{QA}$ changes to a 1. The 0 to 1 transition of $\overline{QA}$ triggers flip-flop B. DB is being fed a 0 from $\overline{QB}$, so QB becomes 0 and $\overline{QB}$ becomes 1. Since $\overline{QB}$ also has made a 0 to 1 transition on this pulse, flip-flop C is also triggered. QC changes to a 1, and $\overline{QC}$ changes to a 0. Reading the outputs from right to left, we now have a binary count of 100, or four.

**Clock Pulse #5**    This pulse again behaves like Clock Pulse #1, changing QA to a 1 and $\overline{QA}$ to a 0, and leaving flip-flops B and C unchanged. The outputs now read 101, or five.

**Clock Pulse #6**    QA changes from 1 to 0, and $\overline{QA}$ changes from 0 to 1. This triggers flip-flop B, causing QB to go from 0 to 1, and $\overline{QB}$ goes from 1 to 0. Flip-flop C is not triggered this time, so its outputs states remain the same. The binary count is now 110, or six.

**Clock Pulse #7**    This pulse behaves like all of the other odd-numbered clock pulses, changing QA to a 1 and $\overline{QA}$ to a 0, and leaving flip-flops B and C unchanged. The outputs now read 111, or seven.

**Clock Pulse #8**    QA changes from 1 to 0, and $\overline{QA}$ changes from 0 to 1, triggering flip-flop B. QB changes from 1 to 0, and $\overline{QB}$ changes from 0 to 1, triggering flip-flop C. QC also changes from 1 to 0, and $\overline{QC}$ goes from 0 to 1. The binary count after the eighth clock pulse is 000 once more. The counter circuit is reset for another series of count pulses.

This pattern will continue repeating as long as there are incoming clock pulses and power is continuously applied to the circuit. You can reset the counter forceably to 000 at any time by pulling the preclear inputs LOW via the SPDT switch.

Although only three stages are shown here, the same principles can be expanded for four or more stages. Most practical counter circuits, especially for large numbers of stages, are in the form of a dedicated IC, but internally they usually function in pretty much the same manner as discussed here.

# Tri-state logic

Normally in digital circuits, all signals must be in one of two definite states. A digital signal is either HIGH (1) or LOW (0), and there are usually no other possibilities. Some circuits do use a special type of digital gate that uses tri-state logic.

As the name suggests, the output of a tri-state gate can have any of three possible conditions: the usual HIGH (1) and LOW (0) digital states, plus a third, high-impedance (Z) state. When in the high-impedance state, the gate electrically looks like an electrical circuit. In effect, its input(s) is isolated from the circuit. It is sort of a switchable digital gate. Since a digital gate is a form of switching circuit in the first place, a tri-state gate is a "switchable switch," which certainly sounds weird but can be highly useful in a number of applications.

In the high-impedance state, the tri-state gate is effectively removed from the circuit. As a result, it is possible to reconfigure a digital circuit under electrical control. Such a system obviously can be made very versatile.

A tri-state gate can duplicate the function of any of the standard digital gates described in the first part of this chapter, but in practice, most tri-state devices are buffers. A tri-state buffer is shown in Fig. 10-17. Notice that, in addition to the normal data input (which we will call "D"), there is a second input, which is used to control the high-impedance output state. We will call this control input "C." A standard HIGH/LOW digital signal on this pin turns the high-impedance function on and off. The truth table for a typical tri-state buffer looks like this:

| Inputs | | Output |
|---|---|---|
| C | D | |
| 0 | x | Z |
| 1 | 0 | 0 |
| 1 | 1 | 1 |

In this truth table, the Z represents the special high-impedance output state. The x means "don't care." The data input is ignored when the control input is LOW, turning on the high-imped-

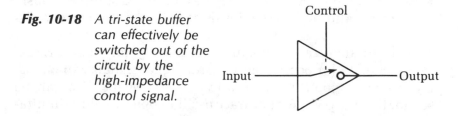

**Fig. 10-17**  *A tri-state buffer's output might be HIGH, LOW, or high-impedance (off).*

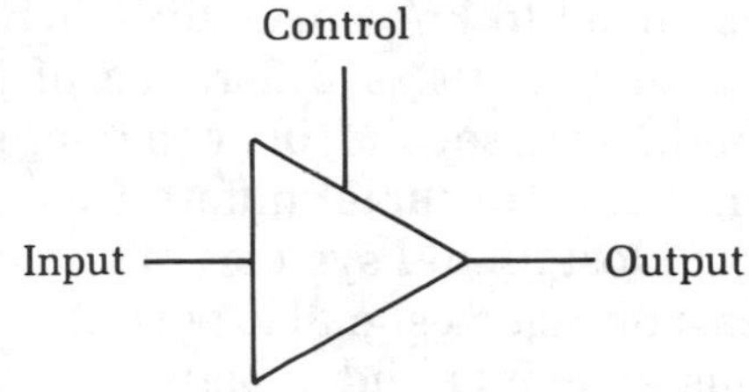

ance output state. When the control input is HIGH, the device acts just like a regular digital buffer — the output state is that same as the (data) input state.

In effect, we have a switchable digital buffer. In some schematic diagrams, a tri-state buffer is shown as in Fig. 10-18 to emphasize this switchlike functioning. When the control input turns on the high-impedance state, the buffer electrically looks like an open switch. The data input is isolated from the circuit and has no effect (unless, of course, a parallel path before the tri-state buffer also routes the same data signal to some other point in the overall circuit).

**Fig. 10-18**  *A tri-state buffer can effectively be switched out of the circuit by the high-impedance control signal.*

It is vitally important to be aware of tri-state logic when you are attempting to service digital circuits. When you are expecting only standard HIGH or LOW digital signals, a high-impedance output can cause major confusion. You could mistake this perfectly normal condition for a serious circuit defect and waste time attempting to "repair" it.

# Sequencing problems

It isn't too hard to determine what is going on throughout a static digital circuit with unchanging signals. You can measure each test point individually and determine whether or not it is at the correct logic level. Most real-world digital circuits aren't quite so cooperative, however. There is usually some sort of sequential element to the digital signals throughout the circuit. In other words, you don't only need to know the logic state at each test point, you also need to know exactly when the signals occurred. In many digital circuits, a difference of just a few milliseconds (thousandths of a second) or even microseconds (millionths of a second) can make a crucial difference.

Most digital systems of any degree of sophistication have at least one clock signal to permit synchronization between the various elements and subcircuits within the system. Sometimes a multiphase clock is used, in which case not all devices or subcircuits occur at the same point in the cycle, but some may be delayed or sped up a partial cycle. Some digital systems use multiple clocks with completely different frequencies for different purposes. In some cases even these disseparate clock signals must be synchronized with one another, while in other instances, they might be free-running with respect to each other.

In addition to clock synchronization, in many digital circuits one event must follow another in a specific sequence. Once again, the timing of all the relevant signals is absolutely crucial.

In any sequential digital circuit, measuring a single test point with a voltmeter or a logic probe just won't give you adequate information. Several test points must be monitored simultaneously to determine the timing relationships between the various signals.

A logic analyzer is designed specifically for this purpose. Usually 8 or 16 (or occasionally more) digital signals can be displayed simultaneously. A multitrace oscilloscope also can be used for this purpose. A dual-trace oscilloscope can only simultaneously display two signals at a time. There are sometimes ways to increase the number of simultaneous signal displays with some external added circuitry. Digital oscilloscopes often can be operated in a logic analyzer mode.

Figure 10-19 shows a typical timing graph for a digital circuit, detailing when each signal should occur with respect to the others. It is often helpful, when you are working on a digital circuit, to make up your own timing chart based on your readings. The correct timing relationships usually will be included in the service data for the equipment. If not, you might have to try to figure out the logical sequence of events (digital signals) to achieve the desired purpose of the circuit in question.

Figure 10-20 shows a timing graph from actual readings on the same circuit whose nominal timing graphs was shown in Fig. 10-19. Notice that signal C is slightly off, signal E is completely incorrect, and signal F is missing altogether. This circuit would be likely to give very erratic operation, if it worked at all.

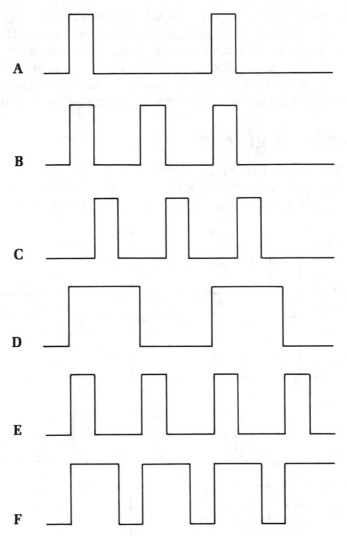

**Fig. 10-19**  *A timing graph for a typical digital circuit.*

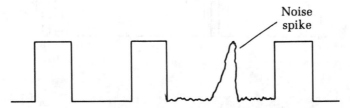

Noise
spike

**Fig. 10-20**  *This measured timing chart indicates sequencing problems when compared to the correct timing chart for the circuit shown in Fig. 10-19.*

Compare these test results with the schematic diagram for the circuit. Devices near the incorrect test points are the most likely suspects in troubleshooting the defect. In this example, we would strongly suspect the circuitry closest to test point E, but it is entirely possible that the small error in signal C is what is somehow throwing signal E so far off from its nominal operation.

# Noise and glitches

Many digital circuits are quite sensitive to errors occurring from noise. A noise spike can make a 0 look like a 1, or vice versa, as illustrated in Fig. 10-21. Such noise spikes can create logic change

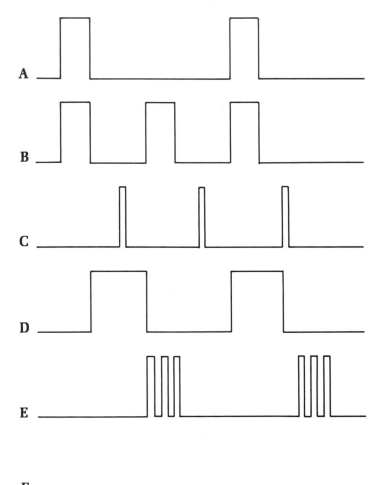

**Fig. 10-21**  *A noise spike might fool the digital circuit into thinking it is a logic signal.*

transitions when they shouldn't occur, causing many potential triggering and synchronization problems throughout the circuit.

Sometimes just a single noise spike will somehow get into the circuit, causing a brief malfunction, which might merely be a momentary oddity or might cause major grief—such as completely clearing the system's memory. Usually no lasting damage to the equipment, if not to the data, will be done by such a glitch. If this just happens once and then the circuit goes back to behaving normally and reliably, it was probably just a stray glitch. It would be impossible to accurately troubleshoot such a problem because you are of necessity working after the effect, and the noise spike that caused the glitch is long gone, so there is no way to measure it.

Noise spikes like this, whether stray glitches or frequent occurrences, most often get into the circuit through the power supply. A good power line filter and/or a surge suppressor often will clear up noise-related problems.

Internal circuit filtering is also helpful. Ideally, every digital chip in the entire circuit should have its own filter capacitor. This capacitor does not need a large value. Something in the 0.001 $\mu$F to 0.01 $\mu$F range usually will be suitable. Connect this capacitor between the power supply pins (V + and ground) as physically close to the body of the protected IC as possible. You certainly don't want the lead to the filter capacitor to act as an antenna for noise signals through RF pick-up. If the filter capacitor is an inch or two away from the IC it is supposed to protect, it won't be able to do its job. Physical placement of such filter capacitors is very crucial.

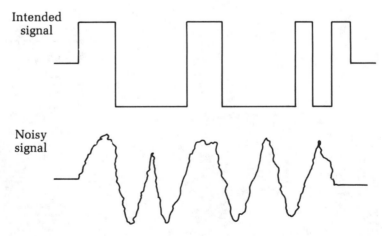

**Fig. 10-22**  *Desired logic signals can be partially or completely obscured by picked up noise.*

Noise in signal lines is usually a result of RF pick-up in one form or another. This noise can cause a digital circuit to misfunction in a number of strange ways because the logic transitions in the desired data is partially or completely obscured by the inadvertently picked up noise signal(s), as illustrated in Fig. 10-22. Often increased shielding of the appropriate signal lines will clear up such problems. In some cases, physically moving the entire circuit might remove it from a particularly strong RF energy field that is inducing noise into the data lines.

# 11
# Flowcharting and troubleshooting

IF YOU ARE SERVICING AN ELECTRONIC CIRCUIT WITH JUST A DOZEN components or so, you can take a brute-force approach and just test everything. This method would, however, be highly impractical for any circuit with any degree of complexity. A television set, a computer, or a VCR (to suggest just a few examples) might contain hundreds, perhaps even thousands, of components. It would not be at all reasonable to attempt to test everything in such a device.

If you are attempting to service a given piece of equipment, it must be exhibiting symptoms of some sort. That is, it is presumably misfunctioning in some way. The misfunction might be simple (a specific indicator light isn't going on), or it might be global (the entire device doesn't work at all). Often, there are multiple symptoms that might or might not be related (for example, a TV set might have no sound and a compressed picture). Whatever the symptoms are, they always provide important clues for servicing the equipment in question.

You, as an electronics technician, must think about the symptoms logically and try to imagine what could possibly be causing the observed symptoms. In this way you will know what types of tests are most likely to provide worthwhile results. Often, many different causes could be behind a given symptom or group of symptoms, so, if possible, try to list these potential causes in order of probability. Make the most probable tests first; you are less likely to waste time with fruitless tests that way.

When there are multiple symptoms, they might be caused by entirely separate causes, but it is relatively unlikely for a circuit to simultaneously develop two (or more) entirely unrelated defects. It can happen, but it is more likely that all the observed symptoms stem, directly or indirectly, from a single root cause. Use logical thinking to determine what could be causing all the observed symptoms and test for that first. Only if you run into a dead-end on the common cause should you start troubleshooting the observed symptoms independently. Again, this will save you a lot of wasted time in the long run.

# Basic flowchart

A very useful troubleshooting technique that is often ignored is the *flowchart*, a simple block diagram graph of the equipment to be serviced. The functions of the various subcircuits and their interconnections are shown on a flowchart, helping you determine what subcircuit(s) is most likely to be causing the observed symptoms. A very simple flowchart for a typical color television set is shown in Fig. 11-1.

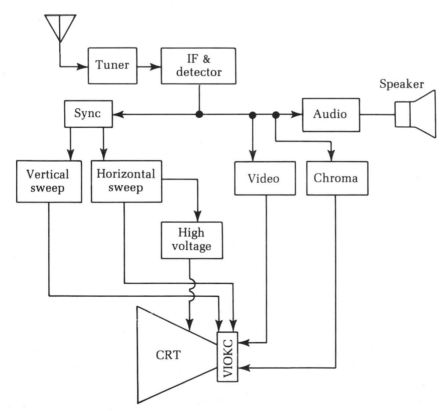

**Fig. 11-1** *A functional flowchart is a very useful troubleshooting tool.*

If you are an experienced technician, you might be able to hold a full flowchart entirely in your head, but for most of us, actually drawing out a flowchart will be helpful, especially in more difficult servicing jobs. There is no reason to be concerned about precise dimensions or perfectly straight lines. Your flowchart is not for publication; it is for your own convenience. If it is neat enough for you to read it, it's good enough. It often helps just to see the circuit's functions broken down in visual form.

As a rule of thumb, it is a good idea to always consider the power supply as a possible culprit for virtually all symptoms. Incorrect supply voltages or noise on the power lines can cause a lot of surprising results in many circuits.

The shapes used in your flowchart don't really matter, although I find it useful to use different shapes for certain common types of subcircuits. For example, using a circle for a signal source (such as an oscillator) or a triangle for an amplification stage can make the flowchart easier to read at a glance.

Often, each section in the original flowchart will include dozens of components, and often can be broken down to multiple subcircuits. It usually will be helpful to break down any suspected section into more detailed subsections. For example, if you are working on a television set with problems in its sound, but the picture is okay, you'll probably only want to look closer at the audio amplifier section. Draw a more detailed sub-flowchart for this section of circuitry, as illustrated in Fig. 11-2. You don't need to identify the subcircuits in any flowchart section where no defect is suspected or likely. You don't expect to do any testing in that part of the equipment's circuitry.

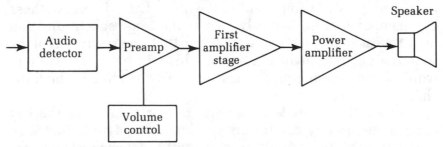

**Fig. 11-2**  *When you've narrowed the troubleshooting down to a specific section, you might want to draw up a more detailed partial flowchart.*

The next step in creating a troubleshooting flowchart is to identify the key components in any relevant stages, especially ICs, transistors, transformers, relays, etc. This technique will help focus you in on which components and connections you will want to test and which probably aren't worth bothering with, since they are unlikely to be defective, given the symptoms observed. For example, in a TV set with sound problems (and no other symptoms), there would be little point in testing the picture tube or the color burst oscillator.

Write the appropriate component identification in flowchart, as illustrated in Fig. 11-3. It will be easiest to do so by comparing your flowchart with the schematic included in the servicing data. It is very difficult to service a complex electronic circuit without a schematic unless you have a great deal of experience with that particular type of equipment.

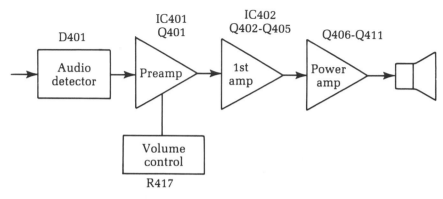

**Fig. 11-3**  *It is a good idea to write the key components for each stage into the appropriate spaces on the flowchart.*

You don't have to list every single resistor or capacitor. Just list the key components, especially active components, and controls. Any passive component in the immediate vicinity of these key components is obviously part of the same subcircuit. You might occasionally test a resistor or two that isn't really part of the subcircuit you thought it was, but this will be fairly rare and won't waste as much time as getting overly detailed in the flowchart.

Notice that there is little point in worrying about the key components in any subcircuits you have already eliminated in the troubleshooting process. If you know the defect must be somewhere in the audio amplifier of the TV, the video amplifier and the tuner are probably okay, so you don't care about locating their key components.

A little logic and a good flowchart will tell you what tests you should perform to track down the defect(s) causing the observed symptoms.

# Servicing without a schematic

All too often it is necessary to service a piece of electronic equipment for which no schematic diagram or other relevant technical literature is available. Clearly this problem increases in seriousness with the complexity of the equipment in question.

The key to solving such problems is to first mentally break up the circuitry into functional blocks. For moderate to complex equipment, it is probably a good idea to actually draw out a block diagram, although, for some relatively simple circuits, you might be able to hold the information in your head. Of course, skill in this type of task can come only with experience. While it will get easier as you gain practical servicing experience and theoretical knowledge, it will never be truly easy. As a result, you should first make every attempt to find a schematic or other service literature if at all possible.

In many cases, there will be some things you simply will not be able to figure out on your own, so if you don't have a schematic, you might be out of luck on some special features and calibration procedures, and the like. Still, there is much you can do with most electronic equipment without a schematic. Just remember, this approach is one of last resort. A schematic will make your job 100 percent simpler.

It is difficult to look at a section of circuitry and determine what it is supposed to do, particularly when one or more exotic, unfamiliar, or unmarked ICs are used. An amplifier IC looks just like a timer IC, which looks just like a digital flip-flop IC. Obviously, you won't get very far with your block diagram from just staring at the circuit board. You have to use some logic.

Do not start from the circuit itself, but from the equipment's intended function. What circuit stages would be required to do the job? In other words, you have to think like a circuit designer does. Make a note of all controls on the equipment. The controls will provide valuable clues to circuitry functions. Also, components near the control are presumably part of the same subcircuit.

Don't forget "obvious" stages. Almost any piece of electronic equipment will have a power-supply subcircuit of some sort. For some portable equipment, this circuit will be nothing more than some batteries and perhaps a diode or two to protect against possible incorrect installation of the batteries. All you really have to check is the battery voltage and the condition of the diodes, mainly to ensure that they are not shorted out.

Ac-powered equipment generally will have a more complex power-supply stage. First check for blown fuses or circuit breakers. (Of course, you should make sure the equipment is plugged in.)

Most ac power supplies have a power transformer of some sort. The secondary winding (or windings) of this transformer will be connected to either one or more rectifier diodes or a voltage regulator IC. (In some circuits both might be encountered). In either case, there probably will be some filter capacitors, too. Usually at least some of these filter capacitors will be fairly large electrolytics, which not only have large capacitance values, but also are physically large. The transformer and the large filter capacitors usually make it relatively easy to identify and isolate the power-supply circuitry.

A voltage-regulator IC usually resembles a slightly oversized power transistor with three leads (in, out, and common). Sometimes a bridge rectifier is used instead of four separate rectifier diodes. This component will look like a square block of plastic with four leads. Internally, it is simply four matched rectifier diodes encapsulated in a single convenient housing.

Locating the power supply also will provide you with valuable clues in tracing the remainder of the circuitry. Find the ground or common connections. Identifying the supply voltage and common connections to any ICs will help cut down the possibilities for identifying the inputs and outputs.

In most electronic equipment, some sort of signal probably will pass through the system. Attempt to locate the point of origin for the main signal. If the equipment uses some sort of external signal source, this should be a relatively easy task. Just find the input jacks or other connectors. For example, an amplifier in a stereo system might take its input signal from a tape deck or a CD player. In a TV or a radio, the main signal originates at the antenna terminals. In other equipment, the main signal might be generated internally, more often than not, the work of an oscillator stage of some sort. An oscilloscope would be a big help in locating a signal's point of origin and tracing its passage through the circuitry.

If the equipment you are working on uses signals in the audible range, you might be able to use a signal tracer probe, which can be a simple audio amplifier and speaker with a suitable probe at its input. Refer to chapter 9 for more information on signal tracing.

Pay close attention to any controls, since they will give you clues as to the function of that section of the circuitry. For example, a volume control or gain control is almost certainly part of an amplifier stage of some sort. (Always remember, that stage might perform additional functions in addition to simple amplification.)

Servicing electronic equipment without a schematic is not easy and should be avoided if at all possible. Sometimes, however, it is necessary. Just remember, the task is not truly impossible. Work as slowly and logically as possible. Take as many measurements as you can. It is hard to tell if any single measurement is correct or not when you look at it by itself. If, however, you take many measurements throughout a circuit, one or two might stick out like a proverbial sore thumb.

Try to relate the odd-seeming measurement(s) with the symptoms exhibited by the malfunctioning equipment. If, for example, the problem is excessive distortion — if one stage of the circuit suddenly shows a lot of mysterious and seemingly unnecessary signals that might possibly be the source of the distortion. Try to think if there could be any possible reason for the "odd" signals to be part of the intended design of the circuit. If not, what would have to be done to eliminate them? Does that clear up the misfunction, or does it create new problems?

For obvious reasons, servicing without a schematic or other relevant technical literature or data is not at all recommended for inexperienced electronics technicians. To get some experience, try your hands on some non-crucial equipment, especially equipment someone else has thrown out because of some defect.

Another way to learn to do such "data-blind" servicing would be to first attempt diagnosis on a piece of equipment before you look at the schematic and other troubleshooting data. Don't actually do anything to the circuitry at this point — just attempt to figure out how it is supposed to work. Then refer to the available servicing data and see how close you came. The first few times you try this method, you almost certainly will be way off more often than not, but you will become more skilled with time.

Remember, servicing electronic equipment without a schematic or other appropriate servicing data is always a course of last resort. Only a very foolish technician would ever consider attempting a repair without consulting any and all schematics or troubleshooting materials. The only possible exception would be if you have worked on one particular model of equipment so often that you have the relevant servicing data memorized. Even then, it is probably unnecessarily risky to rely too heavily on human memory. Keep the schematic and other materials handy so you glance at them occasionally, just to be sure.

# Systems comprised of multiple pieces of interconnected equipment

In large electronics systems, multiple pieces of equipment are interconnected and functionally interact. One obvious example is a computer system, which might include the computer itself, a disc drive, a monitor, a printer, a mouse, and perhaps a few other peripherals. Another example is a stereo system made up of an amplifier, an FM tuner, a tape deck, an equalizer, and a CD player.

When a malfunction shows up in such a system, you obviously need to track down the fault to the appropriate device so it can be serviced. It would be a waste of time to attempt to service the stereo amplifier when the problem is in the equalizer.

The flowchart method described earlier in this chapter is idea for troubleshooting such systems. Each device in the system is given its own functional block in the system flowchart. Figure 11-4 shows a flowchart for the computer system. Our typical stereo system is flowcharted in Fig. 11-5.

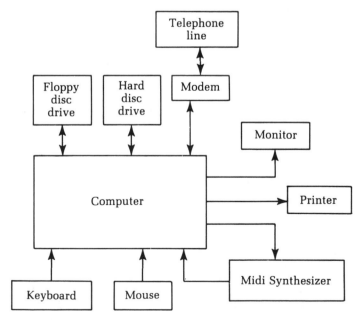

**Fig. 11-4**    *A flowchart also can be used for multiple devices connected into an interactive system, such as a computer system.*

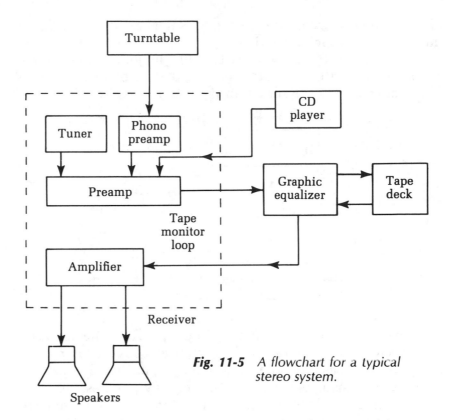

**Fig. 11-5**  *A flowchart for a typical stereo system.*

## Computers

The field of electronics is constantly expanding and growing, and new types of devices keep turning up. Sooner or later almost any technician will have to service some sort of unfamiliar type of electronic equipment. This can seem to be a very intimidating task. Remember that the basic principles of electronics do not change; they are simply put together in new ways.

Once again, we strongly recommend flowcharting the unfamiliar equipment you are attempting to service. What functional stages must it have to do its job?

Personal computers seem to be everywhere these days. Although their prices have come down incredibly in the last decade, a defective computer is still usually too expensive to be considered disposable. Since few people are willing to junk and replace an old computer when it starts misfunctioning, someone will have to service these devices.

First off, remember that a computer is a digital device, and except for the power supply and portions of the monitor, it probably uses digital circuitry exclusively. Refer to chapter 10 for an introduction to digital servicing.

A computer typically consists of five major sections: power supply, clock, CPU, memory, and I/O port. A block diagram of this basic computer is shown in Fig. 11-6.

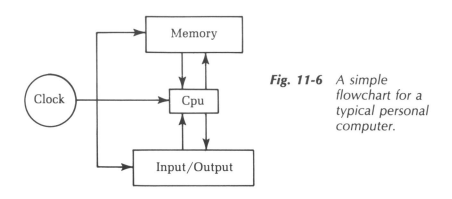

**Fig. 11-6** *A simple flowchart for a typical personal computer.*

The power supply, of course, simply provides the necessary voltages to operate the rest of the computer's circuits. It is essentially the same as any other power supply, except that particularly tight filtering is usually required. Any glitches, ripple, or noise in the supply voltages can confuse the data and produce incorrect results. In some cases, portions of the computer might flatly refuse to operate at all. The computer will "lock up."

The clock is the source of the master synchronization control signal, used to keep all the various subcircuits in the computer operating together with the proper timing. Each signal must be transmitted and received at precisely the correct instant or nothing will function properly. The clock is basically just a very precise oscillator circuit, typically operating at a frequency of at least several megahertz (1 megahertz = 1,000,000 Hertz).

For maximum frequency precision, the clock in a computer is almost always crystal-controlled. If the clock is defective, various signals in the computer will be mistimed, resulting either in total garbage at the output or locking up of the system altogether.

The CPU is the central processing unit — the "brain" of the computer. This circuitry performs all the various mathematical and logical operations on the data. It routes data to and from the memory and/or the I/O port. In a practical personal computer, the CPU is virtually always in the form of a large IC, usually with about 40 pins. The CPU chip is usually fairly easy to locate on the computer's printed circuit board — it will almost always be the physically largest IC in the circuit.

If the CPU goes bad, all sorts of strange things can happen. The output data will be more or less random garbage, the computer might lock up, or the entire system might simply shut itself down. This last could look like a power-supply problem.

The memory section of a computer is used to store data and program instructions. There are two basic types of memory: ROM and RAM. ROM stands for *read-only memory*. The digital data stored in ROM is permanent and cannot be changed. The computer's operating system, for example, is stored in ROM. The data stored in ROM is not lost or otherwise affected when power to the computer is interrupted.

RAM is short for *random access memory*, meaning any piece of data can be independently accessed, without sequentially stepping through all the preceding or proceeding memory locations. Actually, this is not a very appropriate name since ROM is also randomly accessible. Some technicians prefer to call RAM *RWM*, for Read-Write Memory. This is a much more descriptive name. The CPU can read data from or write data to RAM.

This type of memory is used for temporary storage, such as the instructions for a particular program and the appropriate data. For example, we are writing this book on our computer. The word processing program and the text we have written is stored in RAM. The data in RAM is erasable. For most types of RAM, an interruption in the computer's power source will clear out all stored data. Some newer-technology RAM devices are more or less nonvolatile and can store data even if the system's power is interrupted.

Defects in memory are usually localized. Only some of the data will be disrupted. For example, a program might work just fine until it reaches the instructions stored in the defective section of memory. In other cases, just a certain section of stored data might be affected.

Data is carried between the CPU and the memory on *buses*. A bus in this case is a multiline path for digital data. CPUs and memories are designed to group digital bits into words. Some early computers used four-bit words called *nybbles*, but the eight-bit *byte* is the nominal standard for personal computers. Today 16- and 32-bit computers are becoming more common. In some cases, an increased word size will result in faster operation of the computer.

The bus consists of one signal line per bit in the standard digital word used by that particular computer. Some buses are one way, such as the bus from ROM to the CPU. Other buses can transfer data signals in either direction, such as the bus between the CPU and RAM. A defect in a bus will affect all data passing through it. Usually, only one bit will be affected, distorting the data in a fairly predictable way. For example, bit 3 might always be LOW or it might always be HIGH.

The I/O (input/output) port permits the CPU to communicate with the outside world. Unless you can get data into and out of the computer, it isn't going to do anyone any good at all. Typical computer input devices include the keyboard, a mouse, a joystick, or a sensor of some sort. Standard computer output devices include the monitor (a TV-like display screen) and a printer. Some I/O devices work both ways, providing input to the computer and accepting output from the computer. A typical example is the modem, which permits the computer to communicate with a second computer over telephone lines. Another example is a MIDI (Musical Instrument Digital Interface), used with music synthesizers. External memory devices such as floppy disc drives are also full input/output devices. The data stored on the disc is fed into the computer and stored in RAM, then the output data is fed out to the disc drive and recorded onto the disc.

Note that some of these "external" I/O devices might actually be built into the same case as the computer itself. For example, the keyboard is almost always built in, and some computers have a built-in floppy disc drive. However, even these built-in peripherals are still considered "external" to the main computer circuitry itself.

A more detailed flowchart for a typical computer system is shown in Fig. 11-7. Notice that more specific devices are included. They are still part of the original groupings. We are just showing the subsections in addition to the main sections of the system.

## VCRs

The video cassette recorder (VCR) has been one of the most remarkable successes in electronics marketing in the last few decades. The ability to record and play back TV shows and feature films is obviously appealing to vast numbers of consumers. Millions of VCRs are now in use, which means that there is a strong need for VCR service personnel.

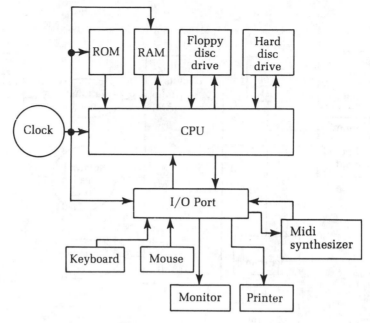

**Fig. 11-7**  *A more detailed flowchart for a typical personal computer.*

It would take an entire book to even begin to adequately discuss servicing and repairing VCRs. In this chapter, we can just take a very brief glance at a few of the basics involved.

Except for playback-only machines and camcorders (a VCR built into a video camera), VCRs include a complete television tuner. This circuitry is essentially the same as the tuner stages in an ordinary television receiver and is serviced in the same way.

Today, virtually all home VCRs are of the VHS type, although some Beta VCRs are still in use. The differences between these two systems are relatively minor when it comes to servicing the equipment.

Figure 11-8 shows a simplified block diagram for a typical VCR. There might be slight variations in some models. Some deluxe VCRs might have a few additional stages. As a general rule of thumb, however, this block diagram is a good starting point when you are troubleshooting almost any VCR.

**Head-switching frequency**  The main synchronization signal in a VCR is usually the head-switching frequency. Multiple record/playback heads are mounted on a rotating drum. These heads are electrically switched in and out of the circuit at appropriate times so the head currently in contact with the tape is activated. The standard for VCR head-switching frequency is 30 Hz.

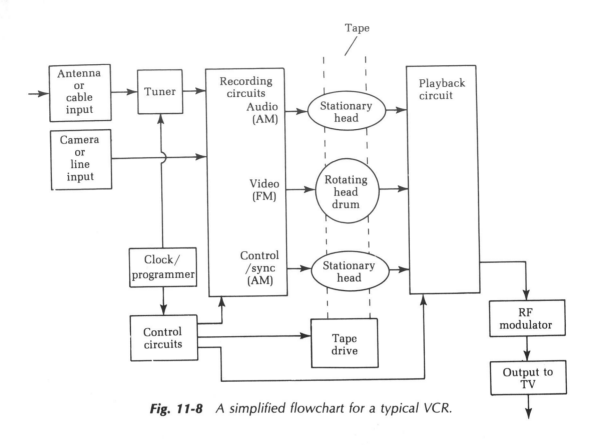

**Fig. 11-8** *A simplified flowchart for a typical VCR.*

The clock signal is in the form of a pulse, which can be used in either analog or digital circuitry. In most VCRs, analog circuitry is used for the actual tuner and record/playback electronics, while digital circuitry is commonly (though not invariably) employed for switching, control, and display functions.

The 30 Hz head switching pulse signal is used as a master synchronization signal for many of the circuit stages throughout the VCR. Obviously, this frequency must be correct and clean. You can use an oscilloscope to check this signal. Especially watch out for distortion of the waveform. You should get a clean, sharp pulse. Sloped sides (slew) or ringing (nonflat tops and bottoms) can cause many stages of a VCR to malfunction.

**Mechanical problems**  Unlike a television set, a VCR has many mechanical, moving parts, in addition to the actual electronic circuitry. Many VCR problems are mechanical, rather than electronic. If the tape cassette will not load or eject, or if the tape will not run, there might be a problem in the electronic control circuitry, or it might be a mechanical problem. Unless the tape is jammed in the machine, always first try a different tape before drawing any conclusion or attempting any repairs. Often, the problem is in the cassette itself. Discard the bad tape before it can cause more trouble.

Another common problem is a defective load belt. This belt may be just dirty, it might be worn slick (causing it to slip), or it might be dried out and cracked. Of course, the VCR will not function if the load belt is actually broken. This situation is usually very easy to see once you have removed the unit from its case.

When checking out and servicing mechanical parts in a VCR, be very careful and gentle. Never force anything. If a given part exhibits any resistance to moving in a certain way, it either isn't supposed to move that way (at least under the current control conditions), or something is preventing its movement. Attempting to force it will only cause something to bend or break, resulting in a bigger and more expensive repair job. Many of the mechanical parts are extremely small and very difficult to put back into place if you accidentally knock them out of position. In some cases, you won't be able to put some parts back into place without a specialized tool.

The delicate mechanical parts in a VCR must be kept scrupulously clean. A little dirt or other contamination in the wrong place can jam up the works. In some cases, it can lead to permanent damage as the machinery attempts to move against an obstruction and one or more parts get bent.

Regular use of a head-cleaning cassette is vital for any VCR. Regular maintenance internal cleaning is also advisable. Certainly, any time you service a VCR, you should take time to clean it, even if that is not part of the current problem. It will help prevent unnecessary problems in the future.

Use cleaning sticks to reach into all the little nooks and crannies. Cotton-tip cleaning sticks are like extra long Q-tips, and are good for general use. Do not overuse a cotton cleaning stick. Once it gets dirty, it will just smear the dirt around, doing more harm than good. Discard such sticks frequently and use a new one. Certainly, if you see visible dirt on the cotton swab, you should discard it and continue working with a new stick. Even if you don't see any dirt, discard the cotton cleaning stick after cleaning two or three mechanical parts. Some dirt might not be visible but can still be potentially damaging if spread around. On an average, you probably should go through a couple dozen cotton cleaning sticks each time you clean a VCR.

Cotton cleaning sticks can be used either dry or with a cleaning fluid, designed for the purpose. Any impurities in the cleaning fluid can cause damage, so use only a cleaning fluid recommended for use in VCRs. Be careful not to get head-cleaning fluid or any alcohol-based fluid on any rubber parts, as it could cause corrosion. Certain types of plastic also might react negatively to alcohol. When in doubt, use head-cleaning fluid only metallic parts and surfaces.

Some special (and more expensive) cleaning sticks are also available and are good for special purposes. For example, foam-tipped cleaning sticks are very good for cleaning the tape path, including pinch rollers, idlers, and rubber belts. Use foam-tipped cleaning sticks with an appropriate cleaning fluid. Discard them once they get dirty and use a clean one. A foam-tipped cleaning stick will generally stay clean a little longer than a cotton cleaning stick.

Some cleaning sticks have chamois heads. Chamois is a very soft, lint-free cloth. It is very good for cleaning dirty video heads without risking scratches. Chamois cleaning sticks can be used dry or, more commonly, with head-cleaning fluid. They usually can be reused several times before replacement is necessary.

**How video tape is recorded**   Video tape is not recorded like ordinary audio tape. On a standard audio tape, the signal is recorded continuously and linearly, as illustrated in Fig. 11-9A. Figure 11-9B shows how signals are recorded on a video tape. One head records one field as a slanted bar, then the next field is recorded by a different head. Low-cost VCRs have two heads. Deluxe models usually have four heads.

This odd-looking method is used to conserve tape. Video data is much more complex than an audio signal, so it takes up more space on the tape. If straight linear recording (like an audio recording) was used in a VCR, the tape would have to move at an incredibly high speed, and even a very long tape would only record a program of just a few minutes. Such a system would not be economical or practical.

In a VCR, the tape itself moves at a fairly slow speed, but the heads are mounted in a drum, which spins very rapidly. This method allows the machine to cram more recorded signal into considerably less physical space.

Because of the way the video signal is recorded and the delicate precision of the tape path in the VCR, a video tape must never be spliced. Discard a broken or damaged tape. Another reason a

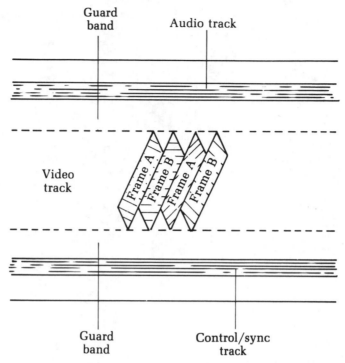

**Fig. 11-9**   *Video data is recorded as angled bars instead of a straight line to conserve tape.*

spliced videotape won't work is the control track recorded along the bottom of the tape. This recorded signal is vitally important for correct synchronization of the video signal when it is played back. The television set will not be able to reconstruct the recorded picture without the continuous signal on the control track.

The audio track along the top records the sound portion of the program. It is pretty much similar to an ordinary audio tape recorder. In stereo VCRs, the audio and control tracks are arranged a little differently, but the basic principle is the same.

Video tapes are recorded along the entire width of the tape, so the cassette cannot be turned over to record on the "other side" like an ordinary audio cassette tape. There is no other side to a video cassette.

# Index